Fortschritte der Chemie organischer Naturstoffe

Progress in the Chemistry of Organic Natural Products

61

Founded by L. Zechmeister
Edited by W. Herz, G. W. Kirby, R. E. Moore,
W. Steglich, and Ch. Tamm

Authors:
D.G.I. Kingston, A.A. Molinero, J.M. Rimoldi

Springer-Verlag
Wien New York 1993

Prof. W. HERZ, Department of Chemistry,
The Florida State University, Tallahassee, Florida, U.S.A.

Prof. G. W. KIRBY, Chemistry Department,
The University, Glasgow, Scotland

Prof. R. E. MOORE, Department of Chemistry,
University of Hawaii at Manoa, Honolulu, Hawaii, U.S.A.

Prof. Dr. W. STEGLICH, Institut für Organische Chemie der Universität
München, München, Federal Republic of Germany

Prof. Dr. CH. TAMM, Institut für Organische Chemie der Universität Basel,
Basel, Switzerland

© 1993 by Springer-Verlag/Wien
Softcover reprint of the hardcover 1st edition 1993
Library of Congress Catalog Card Number AC 39-1015

Typesetting: Macmillan India Ltd., Bangalore-25

With 4 Figures

ISSN 0071-7886
ISBN-13: 978-3-7091-9244-3   e-ISBN-13: 978-3-7091-9242-9
DOI: 10.1007/978-3-7091-9242-9

Contents

List of Contributors   IX

The Taxane Diterpenoids
By D.G.I. KINGSTON, A. A. MOLINERO, and J.M. RIMOLDI   1

1. Introduction.   3
2. General Structural Characteristic and Nomenclature   8
3. The Families of Taxane Diterpenoids   10
   3.1. Taxoids with a C-4(20) Exocyclic Double Bond   11
   3.2. Taxoids with a C-4(20) Epoxide   22
   3.3. Taxoids with an Oxetane Ring.   22
   3.4. Miscellaneous Taxoids   32
4. The Chemistry of the Taxoids   32
   4.1. Isolation Techniques   32
      4.1.1. Extraction.   32
      4.1.2. Purification and Analysis   33
   4.2. Spectroscopy.   35
      4.2.1. UV and ORD/CD Spectroscopy   35
      4.2.2. Infrared Spectroscopy.   37
      4.2.3. ^{1}H-NMR Spectroscopy   37
      4.2.4. ^{13}C-NMR Spectroscopy   39
      4.2.5. Mass Spectrometry   53
      4.2.6. X-ray Crystallography   60
   4.3. Chemical Reactivity   62
      4.3.1. Acylation and Other Protective Group Chemistry.   62
      4.3.2. Hydrolysis   66
      4.3.3. Epimerization at C-7   68
      4.3.4. Oxidation   69
      4.3.5. Reduction   73
      4.3.6. Rearrangements and Related Reactions   76
      4.3.7. Photochemistry   80
5. Approaches to the Synthesis of Taxane Diterpenoids   81
   5.1. Linear Strategies   83
      5.1.1. Biomimetic Approaches   83
         5.1.1.1. Kato's Approach   83
         5.1.1.2. Frejd's Approach   85
         5.1.1.3. Pattenden's Approach   86

     5.1.2. Intramolecular Diels-Alder Approaches 88
         5.1.2.1. Shea's Approach. 88
         5.1.2.2. Jenkin's Approach 90
         5.1.2.3. Sakan's Approach 92
         5.1.2.4. Yadav's Approach 92
     5.1.3. AB → ABC Approaches. 93
         5.1.3.1. Martin's Approach 93
         5.1.3.2. Holton's Approaches: The Synthesis of Taxusin 94
         5.1.3.3. Oishi's Approach 96
         5.1.3.4. Fetizon's Second Approach 99
         5.1.3.5. Blechert's Second Approach. 100
         5.1.3.6. Wender's C-Ring Annulation Approach 101
         5.1.3.7. Kraus' Approach 102
         5.1.3.8. Yamada's Approach 102
         5.1.3.9. Fetizon's Third Approach 103
         5.1.3.10. Gadwood's Approach 104
     5.1.4. BC → ABC Approaches. 105
         5.1.4.1. Swindell's Approach 105
         5.1.4.2. Wender's A-Ring Annulation Approach 108
         5.1.4.3. Sieburth's Approach 110
         5.1.4.4. Kanematsu's Approach 111
 5.2. Convergent Strategies 113
     5.2.1. AC → ABC Approaches. 113
         5.2.1.1. Kitagawa's Approach 114
         5.2.1.2. Fetizon's First Approach. 116
         5.2.1.3. Kende's Approach 118
         5.2.1.4. Funk's Approach 120
         5.2.1.5. Kuwajima's Approach. 120
     5.2.2. A[B]C → ABC Approaches 121
         5.2.2.1. Trost's Approach 121
         5.2.2.2. Inouye's First Approach 123
         5.2.2.3. Blechert's First Approach 125
         5.2.2.4. Clark's Approach 126
         5.2.2.5. Berkowitz's Approach. 127
         5.2.2.6. Inouye's Second Approach 129
         5.2.2.7. Winkler's Approach 131
         5.2.2.8. A Variation on Fetizon's Second Approach 133
         5.2.2.9. Ghosh's Approach 134
         5.2.2.10. Paquette's Approach. 136
         5.2.2.11. Snider's Approach. 140
         5.2.2.12. Zucker's Approach 142
         5.2.2.13. Frejd's Synthesis of a Secotaxoid 142
 5.3. Partial Synthesis of Taxol and Related Compounds 143
     5.3.1. Synthesis of Taxol and Taxol Analogues from
         13-Cinnamoylbaccatin III 143
     5.3.2. Synthesis of Taxol by Acylation of Baccatin III
         with a Pre-formed Side Chain. 145
     5.3.3. Acylation of Baccatin III with β-Lactams or Oxazinones 149
     5.3.4. Synthesis of Taxol Analogues 150
     5.3.5. Synthesis of Oxetane Models 151

6. Biosynthesis and Biotransformation of Taxoids 154
   6.1. Biosynthesis of Taxoids 154
   6.2. Biotransformation of Taxoids 159

7. Bioactivity of Taxol and Other Taxoids 160
   7.1. Toxicity of Taxus Alkaloids 160
   7.2. Biological Activity of Taxol and Related Compounds 161
       7.2.1. Antitumor Activity of Taxol 161
       7.2.2. Microtubule Assembly Activity of Taxol 163
       7.2.3. Structure-activity Relationships of Taxol Analogs 165

Addendum 165

Acknowledgements 173

References 173

Author Index 193

Subject Index 201

List of Contributors

KINGSTON, Prof. D.G.I., Department of Chemistry, Virginia Polytechnic Institute and State University, Blacksburg, Virginia 24061, U.S.A.

MOLINERO, A.A., Department of Chemistry, Virginia Polytechnic Institute and State University, Blacksburg, Virginia 24061, U.S.A.

RIMOLDI, J.M., Department of Chemistry, Virginia Polytechnic Institute and State University, Blacksburg, Virginia 24061, U.S.A.

The Taxane Diterpenoids

D.G.I. Kingston, A.A. Molinero, and J.M. Rimoldi, Department of Chemistry, Virginia Polytechnic Institute and State University, Blacksburg, Virginia, 24061-0212, USA

Contents

1. Introduction.   3

2. General Structural Characteristic and Nomenclature   8

3. The Families of Taxane Diterpenoids   10
   3.1. Taxoids with a C-4(20) Exocyclic Double Bond   11
   3.2. Taxoids with a C-4(20) Epoxide   22
   3.3. Taxoids with an Oxetane Ring.   22
   3.4. Miscellaneous Taxoids   32

4. The Chemistry of the Taxoids   32
   4.1. Isolation Techniques   32
     4.1.1. Extraction.   32
     4.1.2. Purification and Analysis   33
   4.2. Spectroscopy   35
     4.2.1. UV and ORD/CD Spectroscopy   35
     4.2.2. Infrared Spectroscopy.   37
     4.2.3. ^{1}H-NMR Spectroscopy   37
     4.2.4. ^{13}C-NMR Spectroscopy   39
     4.2.5. Mass Spectrometry   53
     4.2.6. X-ray Crystallography   60
   4.3. Chemical Reactivity   62
     4.3.1. Acylation and Other Protective Group Chemistry.   62
     4.3.2. Hydrolysis   66
     4.3.3. Epimerization at C-7   68
     4.3.4. Oxidation   69
     4.3.5. Reduction   73
     4.3.6. Rearrangements and Related Reactions   76
     4.3.7. Photochemistry   80

5. Approaches to the Synthesis of Taxane Diterpenoids   81
   5.1. Linear Strategies   83
     5.1.1. Biomimetic Approaches   83
       5.1.1.1. Kato's Approach   83

          5.1.1.2. Frejd's Approach   85
          5.1.1.3. Pattenden's Approach   86
     5.1.2. Intramolecular Diels-Alder Approaches   88
          5.1.2.1. Shea's Approach   88
          5.1.2.2. Jenkin's Approach   90
          5.1.2.3. Sakan's Approach   92
          5.1.2.4. Yadav's Approach   92
     5.1.3. AB → ABC Approaches   93
          5.1.3.1. Martin's Approach   93
          5.1.3.2. Holton's Approaches: The Synthesis of Taxusin   94
          5.1.3.3. Oishi's Approach   96
          5.1.3.4. Fetizon's Second Approach   99
          5.1.3.5. Blechert's Second Approach   100
          5.1.3.6. Wender's C-Ring Annulation Approach   101
          5.1.3.7. Kraus' Approach   102
          5.1.3.8. Yamada's Approach   102
          5.1.3.9. Fetizon's Third Approach   103
          5.1.3.10. Gadwood's Approach   104
     5.1.4. BC → ABC Approaches   105
          5.1.4.1. Swindell's Approach   105
          5.1.4.2. Wender's A-Ring Annulation Approach   108
          5.1.4.3. Sieburth's Approach   110
          5.1.4.4. Kanematsu's Approach   111
5.2. Convergent Strategies   113
     5.2.1. AC → ABC Approaches   113
          5.2.1.1. Kitagawa's Approach   114
          5.2.1.2. Fetizon's First Approach   116
          5.2.1.3. Kende's Approach   118
          5.2.1.4. Funk's Approach   120
          5.2.1.5. Kuwajima's Approach   120
     5.2.2. A[B]C → ABC Approaches   121
          5.2.2.1. Trost's Approach   121
          5.2.2.2. Inouye's First Approach   123
          5.2.2.3. Blechert's First Approach   125
          5.2.2.4. Clark's Approach   126
          5.2.2.5. Berkowitz's Approach   127
          5.2.2.6. Inouye's Second Approach   129
          5.2.2.7. Winkler's Approach   131
          5.2.2.8. A Variation on Fetizon's Second Approach   133
          5.2.2.9. Ghosh's Approach   134
          5.2.2.10. Paquette's Approach   136
          5.2.2.11. Snider's Approach   140
          5.2.2.12. Zucker's Approach   142
          5.2.2.13. Frejd's Synthesis of a Secotaxoid   142
5.3. Partial Synthesis of Taxol and Related Compounds   143
     5.3.1. Synthesis of Taxol and Taxol Analogues from
          13-Cinnamoylbaccatin III   143
     5.3.2. Synthesis of Taxol by Acylation of Baccatin III
          with a Pre-formed Side Chain   145
     5.3.3. Acylation of Baccatin III with β-Lactams or Oxazinones   149

5.3.4. Synthesis of Taxol Analogues 150
5.3.5. Synthesis of Oxetane Models 151

6. Biosynthesis and Biotransformation of Taxoids 154
6.1. Biosynthesis of Taxoids 154
6.2. Biotransformation of Taxoids 159

7. Bioactivity of Taxol and Other Taxoids 160
7.1. Toxicity of *Taxus* Alkaloids. 160
7.2. Biological Activity of Taxol and Related Compounds. 161
7.2.1. Antitumor Activity of Taxol 161
7.2.2. Microtubule Assembly Activity of Taxol 163
7.2.3. Structure-activity Relationships of Taxol Analogs 165

Addendum 165

Acknowledgements 173

References 173

1. Introduction

The taxane diterpenoids form a unique class of natural products containing the unusual taxane skeleton (**1.1**) or closely related skeletons and occurring in various members of the genus *Taxus* (Taxaceae) and closely related genera. Initial interest in the constituents of *Taxus* species was sparked by the known toxicity of *T. baccata*, or the English yew, since human fatalities due to ingestion of this plant were recorded as long ago as the first century B.C.. Thus Julius Caesar, speaking of his wars against the Gallic tribes, writes "Catuvolcus, who was king of half of the Eburones and had joined Ambiorix in the conspiracy, was now old and weak, unable to endure the hardships of war or flight. He solemnly cursed Ambiorix for instigating the conspiracy, and then poisoned himself with yew, a tree which is very common in Gaul and in Germany" (*27*).

1.1

The first chemical study of a *Taxus* species was carried out by Lucas (*163*) who in 1856 isolated an ill-defined alkaloidal substance which he named taxine. Unfortunately, although "taxine" is almost certainly responsible for a major part of the toxicity of the yew, it proved to be a mixture of related compounds and to be very difficult to work with, due to its instability. Early work on the structure of "taxine" was carried out by German, French, Italian, English, Swiss, and Japanese scientists during the period 1856–1943 (82), but of this work only that of WINTERSTEIN and his colleagues resulted in structure elucidation of a significant fragment of the mixture. These investigators obtained a nitrogenous acid on acidic hydrolysis of "taxine", and determined its structure as 3-(dimethylamino)-3-phenylpropanoic acid, later named "Winterstein's acid" (298, 299).

Work on *Taxus* constituents during the 1940's and 1950's continued to focus on the basic Winterstein esters, presumably because they were thought to be the major active (i.e. toxic) principles and because they were relatively easily isolated (albeit as a crude mixture) by acid extraction. Two separate groups, however, converted the mixed Winterstein esters to the more stable cinnamate esters by β-elimination and thus obtained homogeneous material for the first time. The English group, headed by LYTHGOE, worked with material from *T. baccata* and determined the structure of its major ester as O-cinnamoyltaxicin-I triacetate (**1.2**) (164). The Japanese workers, using *T. cuspidata* (Japanese yew) as their starting material, isolated the related material O-cinnamoyltaxicin-II triacetate, or taxinine (**1.3**) as their major component, and independently determined its structure (151, 187, 279).

1.2 R = OH
1.3 R = H

1.4

1.5 R₁ = Ac, R₂ = H
1.6 R₁ = H, R₂ = Ac

The naturally occurring alkaloid mixture originally designated "taxine" by Lucas was separated into the pure compounds taxines A and B by GRAF and his co-workers during the 1950's, and the structures (**1.4**) and (**1.5**) were assigned to these alkaloids (*83, 84*). Interestingly, taxine A (**1.4**) has a rearranged carbon skeleton, while taxine B has the same ring system as O-cinnamoyltaxicin-I triacetate (**1.2**); its structure has recently been reassigned as (**1.6**) (*70*).

As noted above, the early work on *Taxus* constituents concentrated largely on basic substances and used the relatively harsh extraction conditions of dilute sulfuric acid to obtain the desired components. This process necessarily failed to isolate the neutral materials present in the plant and in addition it probably caused the decomposition of many of the more labile ones. When milder extraction methods were applied, however, many neutral compounds were isolated directly. Thus a nitrogen-free compound, taxinine, in the alcohol extract of needles of Japanese yew was isolated as early as 1925 (*146*), and TAYLOR isolated a substance that he called baccatin from the heartwood of *T. baccata* (*269*). This was later renamed baccatin I (**1.7**) by HALSALL, who also isolated various other baccatins from the same source (*32, 47, 48, 49, 50*). NAKANISHI and his co-workers also isolated various taxinine congeners from the Japanese yew during the same time period (*39, 302*), and Japanese and Scandinavian workers isolated taxusin (*67, 154, 183*).

The isolation of new taxanes and their structure elucidation might have remained an interesting but esoteric branch of natural products chemistry were it not for a discovery by WANI and WALL in the late 1960's. These investigators were searching for anticancer compounds in higher plants, and in 1964 the National Cancer Institute, working with this group, detected cytotoxic activity in an extract of the bark of the western yew, *Taxus brevifolia*. Investigation of this extract was made difficult by the low yield of the active compound, but its structure was finally elucidated and the compound was named taxol (**1.8**) (*285*). The key reaction in the structure elucidation was cleavage of taxol by mild methanolysis to an N-benzoyl-β-phenylisoserine methyl ester and the tetraol (**1.9**); the structure of (**1.9**) was established unambiguously by X-ray crystallography of its 7,10-bisiodoacetate.

The structure of taxol is an unusual one, with the N-benzoyl-β-phenylisoserine group esterifying an oxetane-containing diterpenoid at the C-13 position. The structure of this diterpenoid, now known to be baccatin III (**1.10**), has an interesting history. Baccatin III was first isolated by HALSALL and co-workers in 1966 (*32*), but its structure was not elucidated at that time. A tentative structure, based on chemical and spectroscopic evidence was published in 1970 (*50*); this structure (**1.12**)

1.7

1.8

1.9 R₁ = H, R₂ = OH, R₃ = H
1.10 R₁ = Ac, R₂ = OH, R₃ = H
1.11 R₁ = Ac, R₂ = H, R₃ = OH

1.12

had an opened D-ring. The structure of the related compound baccatin V
(**1.11**) was determined unambiguously by X-ray crystallography in 1970
(*47*). When the structure of the tetraol (**1.9**) from taxol was determined, it
was clear that it was closely related to that of baccatin V, and this led to a
reconsideration also of the structure of baccatin III. The structure of
baccatin III was finally confirmed as (**1.10**) by comparison of its manganese dioxide oxidation product with that obtained from taxol (*49*).

    Initial development of taxol as an anticancer drug was slowed by its
scarcity and insolubility in water, but the pharmacological data that
gradually emerged during the 1970's turned out to be very good, with
taxol showing excellent activity in the B16 melanoma assay. The decision
was thus made by the National Cancer Institute, USA, (NCI) in 1977 to
initiate preclinical development of taxol. Interest in taxol as a clinical
candidate was further enhanced by Horwitz's discovery in 1979 that it
has a unique interaction with tubulin, binding to and stabilizing assembled microtubules (*232*). Taxol entered Phase I clinical trials in 1983
and showed some encouraging activity against the refractory tumors that
are found in these trials. The real breakthrough came in the Phase II
trials, however, which began in the second half of the 1980's and are still
continuing. Significant responses were observed in Phase II trials on

patients with refractory ovarian cancer (*177, 222*), and promising activity has also been shown against breast cancer (*105*) and lung cancer, although the sample size in the latter case is too small to be statistically reliable.

These exciting clinical results have catapulted taxol into the forefront of interest as an anticancer drug, and both the NCI and many academic and industrial laboratories are engaged in studies of its chemistry, pharmacology, and clinical use. For this reason this review will necessarily be selective rather than comprehensive and will focus on the structures and chemistry of the various taxane diterpenoids and on synthetic approaches to taxol and other taxanes. The pharmacology of taxol will be reviewed briefly, but no attempt will be made to be comprehensive in this area.

The taxane diterpenoids have been reviewed on several previous occasions. Among the more comprehensive and accessible reviews are those by LYTHGOE (*164*), MORELLI (*185*), MILLER (*180*), SUFFNESS and CORDELL (*254*), GUERITTE-VOEGELEIN, GUENARD, and POTIER (*87*), KHAN and PARVEEN (*130*), CHEN (*36*), BLECHERT and GUENARD (*17*), SWINDELL (*255*), KINGSTON (*131*) and PAQUETTE (*207*). The review by LYTHGOE describes the initial studies of the Leeds group on the structural characterization of O-cinnamoyltaxicin-I triacetate and the related work by Japanese groups on O-cinnamoyltaxicin-II triacetate (taxinine). The review by MILLER gives structural and spectroscopic data for taxanes isolated prior to 1980, while the review by SUFFNESS and CORDELL gives an excellent summary of the chemistry and also the biological activity of taxol through early 1985. The review by BLECHERT and GUENARD, despite its title, summarizes information on the structure and synthesis of taxane diterpenoids through 1989 and also describes the pharmacology of taxol. The review by KINGSTON discusses the chemistry of taxol in detail, including its structure-activity relationships, while those by SWINDELL and PAQUETTE provide comprehensive surveys of synthetic approaches to the taxane diterpenes. An interesting popular account of the yew has also been published (*98*).

This review combines to some extent the scope of the last four reviews cited. The structures of the known naturally occurring taxane diterpenoids will be described first, together with tabular summaries of NMR data of key compounds. The chemistry of both taxol and other taxanes will then be described and this will be followed by sections describing approaches to the synthesis of taxol and related taxanes. Finally, the biosynthesis and pharmacology of key compounds will be discussed. The literature has been covered as far as possible through March 1992.

2. General Structural Characteristics and Nomenclature

The taxane diterpenoids can all be considered as compounds having the taxane skeleton (**1.1**) or a closely related skeleton. According to IUPAC nomenclature the basic ring system is that of [9.3.1.0.^{3,8}] pentadecene, while Chemical Abstracts names it as a methanobenzocyclodecane derivative; taxol, for example, is benzene propanoic acid, (β-benzoylamino)α-hydroxy-6, 12b-bis(acetyloxy)-12(benzoyloxy)-2a,3,4, 4a,5,6,9,10,11,12,12a,12b-dodecahydro-4, 11-dihydroxy-4α,8,13,13-tetramethyl-5-oxo-7,11-methano-1H cyclodeca[3,4]benz[1,2b] oxet-9-yl ester, 2aR[2aα,4β,4aβ,6β,9α(αR*,βS*),1α,12α,12aα,12bα]. The numbering system used for the taxane skeleton (**1.1**) is that adopted by IUPAC (118) and differs in minor respects from an earlier system (165).

1.1

Two additional nomenclatural matters are appropriately discussed at this point. In the first place, the phrase "taxane diterpenoids" is too cumbersome for general use, and the word taxane alone misleadingly implies that the compounds are alkanes. We thus propose to use the term "taxoid" to describe the *tax*ane diterpen*oids*; this abbreviation has also been used by others. Secondly, the so-called taxane ring system normally contains a double bond, and it will thus be referred to as the taxene ring system where appropriate.

The relative and absolute stereochemistry of the taxoids has been determined by several methods. The first assignments were made by KURONO and his collaborators (152) on the basis of ^{1}H-NMR and optical rotary dispersion data, and the absolute configuration was determined by applying HOREAU'S method to three alcohols derived from taxinine (**1.3**). The application of optical rotary dispersion and circular dichroism to the taxoids will be discussed in more detail in Sect. 4.2.1. Independent proof of structure and absolute chemistry of taxinine was obtained from X-ray crystallography: application of the anomalous dispersion method to 2,5,9,10-tetraacetyl-14-bromotaxinol (**2.1**) confirmed it to have the structure and absolute stereochemistry shown (248, 249). The relative stereochemistry of the related taxoid (**2.2**) was established by X-ray analysis of a rearrangement product (32).

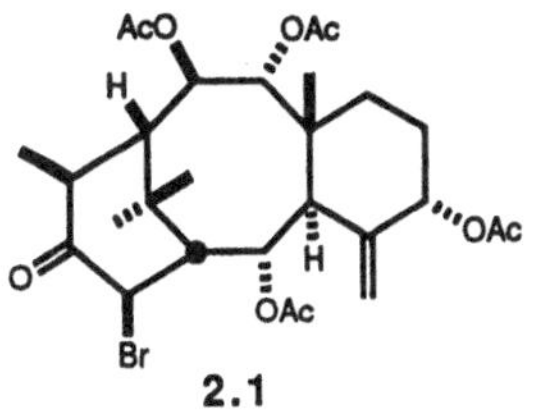

The conventional planar representation of taxol and the other taxoids is, of course, rather misleading. The actual shape of most taxoids, based on the X-ray and other data referred to, is best described as an inverted cup; a stereoview of taxol is shown in Fig. 1. This representation was obtained by using the Chem-3D Plus molecular mechanics program and clearly illustrates the shape of this compound.

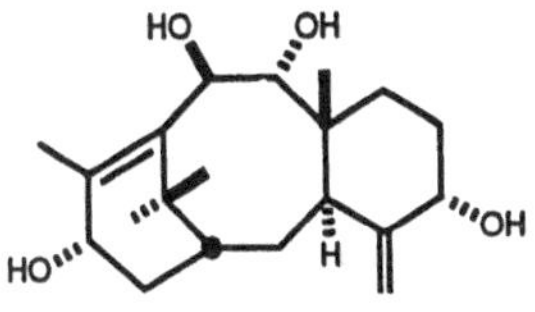

Because of the difficulty of representing taxol and other taxoids in two dimensions, two different representations of the skeleton have been used. The earliest and still the most popular one was used by LYTHGOE and the original Japanese workers and is illustrated for baccatin III as (2.3). This representation is a little easier to draw than the alternate one (which probably accounts for its continued popularity), but it has the disadvantage, as pointed out by LYTHGOE (164), that the orientation of the 16- and 17-methyl groups is the reverse of that intuitively expected.

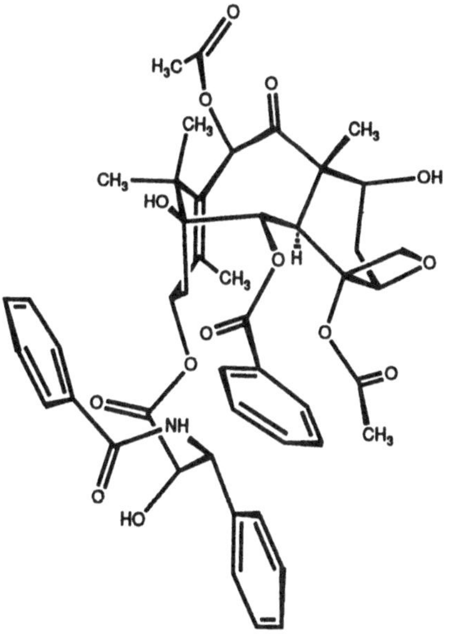

Fig. 1. Stereoview of taxol

The 16-methyl group, which is β with respect to the 19-methyl group as a reference, must be shown with a hatched bond, while the 17-methyl group is α but must be shown with a wedge bond. For this reason we have elected to use an alternate representation which is a modification of one first proposed by MILLER (180); the structure of baccatin III and labeling of the 16- and 17-methyl groups and the 14-hydrogens is shown in (2.4).

2.3   **2.4**

The question of the stability of the taxene ring system has been addressed in a theoretical study using MM2 calculations (257). The key result to emerge from this study is that the Δ^{11} double bond in the taxene ring does not add any additional strain to the system. Thus the taxene hydrocarbon (2.5) has a calculated strain energy of 42.21 kcal/mole, while the corresponding taxane (2.6) has a strain energy of 43.72 kcal/mole, leading to the conclusion that (2.5) is actually *more stable* than (2.6) by about 1.5 kcal/mole. It should be noted that this is not the same thing as saying that the double bond is strain free, since the strain energy of (2.5) is significant and presumably at least a portion of this strain can be associated with the double bond. However, it is clear that the addition of the double bond does not impose any significant additional strain on an already strained molecule.

2.5   **2.6**

3. The Families of Taxane Diterpenoids

The taxoids can be classified in several different ways, depending on the nature of the oxygenation pattern, on the presence or absence of basic side chains, and on other structural features. In this review we have chosen to use the nature of the oxygenation at C-20 and the presence or absence of basic side chains as the major classification criteria.

The primary data for the known taxoids are presented in the following series of extended Tables. Each Table lists the molecular formula and molecular weight for each taxoid, followed by optical rotation, melting point, plant source, plant part, and yield. The yield data are taken directly from the cited references and do not necessarily reflect an optimized isolation procedure; many of the compounds may thus occur in higher yields than those cited. A case in point is 10-deacetylbaccatin III, which is shown in Table 8 as being isolated in yields ranging from $2 \times 10^{-2}\%$ to $1 \times 10^{-4}\%$. However, it has been stated elsewhere (55) that it can be isolated in yields of 0.1% from fresh leaves of *T. baccata*.

In several cases we have made corrections to the published names or structures where these contained an obvious error or for consistency of nomenclature. The code for plant parts is as follows: b = bark, st b = stem bark, hw = heartwood, l = leaves, st = stems, se = seeds, r = roots. Where physical constant or yield data are missing, these data were not included in the cited reference.

3.1. Taxoids with a C-4(20) Exocyclic Double Bond

The most common group of taxoids consists of those compounds with a C-4(20) exocyclic double bond, as exemplified by O-cinnamoyltaxicin-I triacetate (1.2) and other similar taxoids. Simple neutral compounds with an exocyclic methylene group at C-4 are listed in Table 1. All the taxoids in this group carry an α-oxygen substituent (either hydroxyl or acetoxyl) at C-5, and a corresponding β-oxygen substituent at C-10. Most of them are also oxygenated at C-9 and most of them are not oxygenated at C-1.

Basic taxoids with a C-4(20) exocyclic double bond are listed in Table 2. These compounds all carry a Winterstein ester or oxygenated Winterstein ester side chain at C-5, together with other oxygen substituents as indicated.

Taxoids with a 5α-cinnamoyl side chain and a C-4(20) exocyclic double bond are listed in Table 3. Much of the early structural work on the taxoids was carried out on compounds of this type which were obtained as degradation products of the Winterstein ester diterpenoids. However, several members of this class have also been isolated as genuine natural products.

Table 1. *Neutral Taxoids with a C-4(20) Exocyclic Double Bond*

Name	Molecular formula	MW	R_1	R_2	R_3	R_4
5α,9α,10β,13α-Tetraacetoxy-4(20),11-taxadiene (Taxusin) (3.1)	$C_{28}H_{40}O_8$	504	H	H	Ac	H
5α,9α,10β,13α-Tetrahydroxy-4(20),11-taxadiene (3.2)	$C_{20}H_{32}O_4$	336	H	H	H	H
9α,10β-Diacetoxy-5α,13α-dihydroxy-4(20),11-taxadiene (3.3)	$C_{24}H_{36}O_6$	420	H	H	H	H
5α,7β,9α,10β,13α-Pentaacetoxy-4(20),11-taxadiene (3.4)	$C_{30}H_{42}O_{10}$	562	H	H	Ac	OAc
7β,9α,10β-Triacetoxy-2α,5α,13α-trihydroxy-(4)20,11-taxadiene (3.5)	$C_{26}H_{38}O_9$	494	H	OH	H	OAc
2α,5α,Dihydroxy-7β,9α,10β,13α-tetraacetoxy-(4)20,11-taxadiene (3.6)	$C_{28}H_{40}O_{10}$	536	H	OH	H	OAc
2α,5α,9α,10β,13α-Pentaacetoxy-(4)20,11-taxadiene (3.7)	$C_{30}H_{42}O_{10}$	562	H	OAc	Ac	H
5α-Hydroxy-2α,7β,9α,10β,13α-tetraacetoxy-(4)20,11-taxadiene (Decinnamoyltaxinine J) (3.8)	$C_{30}H_{42}O_{11}$	578	H	OAc	H	OAc

R_5	R_6	R_7	$[\alpha]_D$	mp (°C)	Plant source	Plant part; % yield	References
OAc	Ac	OAc	+ 111°	126	*T. baccata*	hw; 1.3×10^{-1}	(*46*)
			+ 120° (CHCl$_3$)	129–131	*T. baccata*	hw	(*67*)
			+ 95° (MeOH)	124–126	*T. mairei*	hw	(*102*)
			+ 168° (CHCl$_3$)	124–126	*T. mairei*	hw; 5.4×10^{-2}	(*161*)
				131–132	*T. cuspidata*	hw; 4.0×10^{-1}	(*183*)
OH	H	OH	+ 134°	195–198	*T. baccata*	hw	(*32*)
OAc	Ac	OH	+ 146°	235	*T. baccata*	hw; 3.8×10^{-4}	(*46*)
			+ 144.2° (CHCl$_3$)	234–236	*T. mairei*	hw; 3.0×10^{-4}	(*161*)
OAc	Ac	OAc	+ 92°	205–207	*T. baccata*	hw; 1.4×10^{-3}	(*46*)
			+ 90.8° (CHCl$_3$)	205–207	*T. mairei*	hw; 2.2×10^{-4}	(*307*)
OAc	Ac	OH	+ 129° (CHCl$_3$)		*Austrotaxus spicata*	b; 7.0×10^{-4}	(*69*)
OAc	Ac	OAc	+ 53° (CHCl$_3$)		*Austrotaxus spicata*	l; 4.4×10^{-4}	(*68*)
OAc	Ac	OAc	+ 146°	165	*T. baccata*	hw; 3.5×10^{-4}	(*46*)
OAc	Ac	OAc	+ 35° (CHCl$_3$)	242–244 (dec)	*T. brevifolia*	b; 2.7×10^{-4}	(*132*)
					Austrotaxus spicata	l; 2.6×10^{-4}	(*68*)

Table 1. (Contd)

Name	Molecular formula	MW	R_1	R_2	R_3	R_4
$2\alpha,5\alpha,7\beta,9\alpha,10\beta,13\alpha$-Hexaacetoxy-(4)20,11-taxadiene (**3.9**)	$C_{32}H_{44}O_{12}$	620	H	OAc	Ac	OAc
2α-α-Methylbutyryloxy-5α-$7\beta,10\beta$-triacetoxy-(4)20,11-taxadiene (**3.10**)	$C_{31}H_{46}O_8$	546	H	OCOiBu	Ac	OAc
5α-Hydroxy-2α-α-methyl butyryloxy-$7\beta,9\alpha,10\beta$-triacetoxy-(4)20,11-taxadiene (**3.11**)	$C_{31}H_{46}O_9$	562	H	OCOiBu	H	OAc
2α-α-Methyl butyryloxy-$5\alpha,7\beta,9\alpha,10\beta$-tetra-acetoxy-(4)20,11-taxadiene (**3.12**)	$C_{33}H_{48}O_{10}$	604	H	OCOiBu	Ac	OAc
$1\beta,2\alpha,5\alpha,9\alpha,10\beta,13\alpha$-Hexahydroxy-(4)20,11-taxadiene (**3.13**)	$C_{20}H_{32}O_6$	368	OH	OH	H	H
2α-Benzoyloxy-$9\alpha,10\beta$-diacetoxy-$1\beta,5\alpha,13\alpha$-trihydroxy-(4)20,11-taxadiene (**3.14**)	$C_{31}H_{40}O_9$	556	OH	$OCOC_6H_5$	H	H
2α-Benzoyloxy-$10\beta,13\alpha$-diacetoxy-$1\beta,5\alpha,9\alpha$-trihydroxy-(4)20,11-taxadiene (**3.15**)	$C_{33}H_{42}O_{10}$	598	OH	$OCOC_6H_5$	H	H
Brevifoliol (**3.16**)	$C_{31}H_{40}O_9$	556	OH	H	H	$OCOC_6H_5$
Taxinine A (**3.17**)	$C_{26}H_{36}O_8$	476	H	OAc	H	H
Taxinine H (**3.18**)	$C_{28}H_{38}O_9$	518	H	OAc	Ac	H

R_5	R_6	R_7	$[\alpha]_D$	mp (°C)	Plant source	Plant part; % yield	References
OAc	Ac	OAc	+ 31°	197	T. baccata	hw; 4.5×10^{-4}	(46)
H	Ac	H	+ 45°	115	T. baccata	hw; 2.0×10^{-3}	(46)
			+ 46.1° (CHCl$_3$)	114–116	T. mairei	hw; 7.7×10^{-5}	(307)
OAc	Ac	H	+ 63°	227–229	T. baccata	hw; 1.2×10^{-3}	(46)
OAc	Ac	H	+ 56°	155–156	T. baccata	hw; 8.0×10^{-4}	(46)
OH	H	OH	− 5.6° (PY:CHCl$_3$) (1:1)	120–121	T. chinensis	st,l	(122)
OAc	Ac	OH	+ 5° (CHCl$_3$)	196–197	T. chinensis	st,l	(122)
OH	Ac	OAc	+ 67.7° (CHCl$_3$)	155–157	T. chinensis	st,l	(122)
OAc	Ac	OH		200–203	T. brevifolia	l; 3.0×10^{-1}	(2)
OAc	Ac	=O.	+ 106° (CHCl$_3$)	254–255	T. cuspidata	l; 8.0×10^{-4}	(39)
				254–255	T. chinensis	l; 4×10^{-4}	(38)
OAc	Ac	=O	+ 96° (CHCl$_3$)	166–167	T. cuspidata	l; 4.0×10^{-5}	(39)

Table 2. *Basic Taxoids with a C-4(20) Exocyclic Double Bond*

Name	Molecular formula	MW	R_1	R_2	R_3	R_4	R_5	R_6	$[\alpha]_D$	mp (°C)	Plant source	Plant part; % yield	References
Austrospicatine (3.19)	$C_{41}H_{55}NO_{12}$	753	H	H	OAc	OAc	OAc	OAc	+52° (CH_2Cl_2)		*Austrotaxus spicata*	l; 5.6×10^{-3}	(68)
2β-Deacetylaustrospicatine (3.20)	$C_{39}H_{53}NO_{11}$	711	H	H	OAc	OAc	OAc	OH	+56° (CH_2Cl_2)	298–300	*Austrotaxus spicata*	l; 3.4×10^{-3}	(68)
7β-Deacetylaustrospicatine (3.21)	$C_{39}H_{53}NO_{11}$	711	H	H	OH	OAc	OAc	OAc	+41° ($CHCl_3$)		*Austrotaxus spicata*	l; 2.2×10^{-4}	(68)
7β,9α-Bisdeacetylaustrospicatine (3.22)	$C_{37}H_{51}NO_{10}$	669	H	H	OH	OH	OAc	OAc	+41° ($CHCl_3$)		*Austrotaxus spicata*	l; 1.1×10^{-3}	(68)
2′β,7β,9α-Trisdeacetylaustrospicatine (3.23)	$C_{35}H_{49}NO_9$	627	H	H	OH	OH	OAc	OH	+41° ($CHCl_3$)		*Austrotaxus spicata*	l; 4.1×10^{-4}	(68)

Compound	Formula	MW	R1	R2	R3	R4	R5	R6	$[\alpha]$	mp	Source	Activity	Ref.
2'β-Deacetoxyaustrospicatine (3.24)	$C_{39}H_{53}NO_{10}$	695	H	H	OAc	OAc	OAc	H	+71° (CH₂Cl₂)		*Austrotaxus spicata*	l; 7.4 × 10⁻³	(68)
2α-Acetoxyaustrospicatine (3.25)	$C_{43}H_{57}NO_{14}$	811	H	OAc	OAc	OAc	OAc	OAc	+37° (CHCl₃)		*Austrotaxus spicata*	l; 1.3 × 10⁻³	(68)
2α-Acetoxy-2'β-deacetylaustrospicatine (3.26)	$C_{41}H_{55}NO_{13}$	769	H	OAc	OAc	OAc	OAc	OH	+19° (CHCl₃)	113–115	*Austrotaxus spicata*	l; 3.1 × 10⁻³	(68)
2α-Hydroxy-2'β-deacetoxyaustrospicatine (3.27)	$C_{39}H_{53}NO_{11}$	711	H	OH	OAc	OAc	OAc	H	+50° (CHCl₃)		*Austrotaxus spicata*	l; 1.6 × 10⁻³	(68)
Comptonine (3.28)	$C_{37}H_{49}NO_{10}$	667	H	H	OAc	OAc	=O	OH	+85° (CHCl₃)		*Austrotaxus spicata*	b; 1.7 × 10⁻⁴	(69)
Taxine B (3.29)	$C_{35}H_{45}NO_{8}$	583	OH	OH	H	OH	=O	H	+116° (CH₃OH)	115	*T. baccata*	1	(70)
									+119° (CHCl₃)	113	*T. baccata*	1	(81, 84)

Table 3. *5-Cinnamoyl Taxoids with a C-4(20) Exocyclic Double Bond*

Name	Molecular formula	MW	R_1	R_2	R_3	R_4	R_5	$[\alpha]_D$	mp (°C)	Plant source	Plant part; % yield	References
5α-Cinnamoyloxy-9α,10β,13α-triacetoxy-taxa-4(20),11-diene (3.30)	$C_{35}H_{44}O_8$	592	H	H	H	Ac	OAc	+118.5° (CHCl₃)	165–166	*T. mairei*	hw	(306)
										T. chinensis	l,s; 2.5×10^{-4}	(311)
5α-Cinnamoyloxy-2α,13α-dihydroxy-9α,10β-diacetoxy-4(20),11-taxadiene (3.31)	$C_{33}H_{42}O_8$	568	H	OH	H	Ac	OH	+7° (CHCl₃)	104–106	*T. chinensis*	l,s; 4×10^{-4}	(311)
5α-Cinnamoyloxy-10β-hydroxy-2α,9α,13α-triacetoxy-taxa-4(20),11-diene (3.32)	$C_{35}H_{44}O_9$	608	H	OAc	H	H	OAc	+29.6° (CHCl₃)	110–112	*T. chinensis*	l,s; 4×10^{-4}	(311)
5α-Cinnamoyloxy-2α,9α,10β,13α-tetraacetoxy-4(20),11-taxadiene (3.33)	$C_{37}H_{46}O_{10}$	650	H	OAc	H	Ac	OAc	+255.3° (CHCl₃)	231–233	*T. mairei*	hw; 3.1×10^{-4}	(307)
2α-Benzoyloxy-5α-cinnamoyloxy-9α,10β-diacetoxy-1β,13α-dihydroxy-4(20),11-taxadiene (3.34)	$C_{40}H_{46}O_{10}$	686	OH	OCOC₆H₅	H	Ac	OH	+6.5° (CHCl₃)	212–214	*T. chinensis*	st,l	(122, 313)

Compound	Formula	MW						$[\alpha]_D$	mp (°C)	Species	Occurrence	Ref.
Taxinine[a] (O-cinnamoyl-taxicin II triacetate) (3.35)	$C_{35}H_{42}O_9$	606	H	OAc	H	Ac	=O	+137° (CHCl$_3$) 128° (CHCl$_3$)	265–267	T. baccata	1	(5, 62, 151)
									266–267	T. chinensis	l; 1×10^{-3}	(38)
									264–265	T. cuspidata	l; 8.8×10^{-2}	(39, 302)
									237–239	T. mairei	hw; 4.0×10^{-4}	(161)
Taxinine E (3.36)	$C_{37}H_{46}O_{10}$	650	H	OAc	H	Ac	OAc			T. cuspidata	1	(302)
Taxinine J (3.37)	$C_{39}H_{48}O_{12}$	708	H	OAc	OAc	Ac	OAc	+36° (CHCl$_3$)	248–249	T. mairei	hw; 2.7×10^{-3}	(182)
									249–251	T. cuspidata	l	(302)
2α-Deacetoxytaxinine J (3.38)	$C_{37}H_{46}O_{10}$	650	H	H	OAc	Ac	OAc	+50° (EtOAc)	248–249	T. mairei	b; 5.0×10^{-3}	(159)
Taxinine B (3.39)	$C_{37}H_{44}O_{11}$	664	H	OAc	OAc	Ac	=O	+93.8° (CHCl$_3$) +84.4° (CHCl$_3$)	171–172	T. mairei	b; 2.0×10^{-2}	(159)
									265–266	T. cuspidata	1	(302)
O-cinnamoyl-taxicin I[a] (3.40)	$C_{29}H_{36}O_7$	496	OH	OH	H	H	=O	+285° (CHCl$_3$)	261–262	T. mairei	hw; 2.0×10^{-4}	(307)
									233–234	T. baccata	1	(5)
O-cinnamoyl-taxicin I triacetate[a] (3.41)	$C_{35}H_{42}O_{10}$	622	OH	OAc	H	Ac	=O	+218° (CHCl$_3$)		T. cuspidata	l; 8.0×10^{-5}	(39)
									237–239	T. baccata	1	(5)

[a] isolated as degradation products of taxine.

Table 4. *Taxoids with a C-4(20) Exocyclic Double Bond and Oxygenation at C-14*

Name	Molecular formula	MW	R_1	R_2	R_3	R_4	R_5	R_6	R_7	$[\alpha]_D$	mp (°C)	Plant source	Plant part; % yield	References
Austrotaxine (3.42)	$C_{41}H_{53}NO_{13}$	767	H	a	OAc	O	OAc	Ac	OAc	$-49°$ (CHCl$_3$)		Austrotaxus spicata	l; 4.8×10^{-4}	(68)
2'-Deacetylaustrotaxine (3.43)	$C_{39}H_{51}NO_{12}$	725	H	a	OAc	O	OAc	Ac	OH	$-41°$ (CHCl$_3$)		Austrotaxus spicata	l; 4.4×10^{-4}	(68)
2'-Deacetoxyaustrotaxine (3.44)	$C_{39}N_{51}NO_{11}$	709	H	a	OAc	O	OAc	Ac	H	$-37°$ (CHCl$_3$)		Austrotaxus spicata	l; 3.3×10^{-4}	(68)
2'β,13α,14β-Trisdeacetylaustrotaxine (3.45)	$C_{35}H_{47}NO_{10}$	641	H	a	OAc	O	OH	H	OH	$-21°$ (CHCl$_3$)		Austrotaxus spicata	b; 8.4×10^{-4}	(69)
7β,9α-Diacetoxy-5α,13α,14β-trihydroxy-10-oxo-(4)20,11-taxadiene (3.46)	$C_{24}H_{34}O_8$	450	H	H	OAc	O	OH	H	—	$+11°$ (CHCl$_3$)		Austrotaxus spicata	b; 6.6×10^{-4}	(69)
Taiwanxan (3.47)	$C_{31}H_{46}O_9$	562	OAc	H	H	β-OAc	H	CO^{i}bu	—	$+43.7°$ (CHCl$_3$)	227–229	T. mairei	hw; 1.7×10^{-4}	(308)
											227–230	T. mairei	hw	(103)

Table 5. *Taxoids with a C-12(16)-Oxido Bridge and a C-4(20) Exocyclic Double Bond*

a = $C_6H_5-CH=CH-CO-$

Name	Molecular formula	MW	R_1	R_2	R_3	$[\alpha]_D$	mp (°C)	Plant source	Plant part; % yield	References
Taxagifine (3.48)	$C_{37}H_{44}O_{13}$	696	Ac	a	H	$-5.4°$ (CHCl$_3$)	265–267	*T. cuspidata*	se; 3.8×10^{-3}	(309)
						$+14.3°$ (CH$_3$OH)	265–267	*T. chinensis*	l,st; 2.5×10^{-3}	(313, 314)
						$+7.5°$ (CH$_3$OH)	265–267	*T. baccata*	l; 5.0×10^{-3}	(34, 35)
2α-Deacetyl-5α-decinnamoyltaxagifine (3.49)	$C_{26}H_{36}O_{11}$	524	H	H	H	$+54.5°$ (CHCl$_3$)	206–207	*T. chinensis*	l,st; 3.5×10^{-4}	(311,312)
5α-Decinnamoyl-taxagifine (3.50)	$C_{28}H_{38}O_{12}$	566	Ac	H	H	$+4.6°$ (CH$_3$OH)	118–120	*T. chinensis*	l,st; 1.7×10^{-3}	(313, 314)
5α-Acetyl-5α-decinnamoyl-taxagifine (3.51)	$C_{30}H_{40}O_{13}$	608	Ac	Ac	H	$-7°$ (CHCl$_3$)	261–263	*T. chinensis*	l,st; 3.5×10^{-4}	(313, 314)
Taxinine M (3.52)	$C_{35}H_{42}O_{14}$	686	Ac	H	OCOC$_6$H$_5$	$-24°$ (CH$_3$OH)		*T. brevifolia*	b	(15)
Taxacin (3.53)	$C_{44}H_{48}O_{15}$	816	Ac	a	OCOC$_6$H$_5$	$-4.8°$ (CHCl$_3$)	285–287	*T. cuspidata*	se; 5.8×10^{-3}	(309)

Several taxoids with a C-4(20) exocyclic double bond and additional oxygenation at C-14 have recently been isolated, and these compounds are listed in Table 4.

A final group of taxoids with the C-4(20) exocyclic double bond has an oxido bridge linking C-12 and C-17. These compounds are listed in Table 5.

3.2. Taxoids with a C-4(20) Epoxide

The second major group of taxoids consists of those compounds with a C-4(20) epoxide. Simple compounds of this type are listed in Table 6. The C-5 position is again always oxygenated, with an α-hydroxyl, an α-acetoxyl, or an α-cinnamoxyl substituent. All compounds of this class isolated to date also carry oxygenation at C-2, C-9, C-10, and C-13.

Basic taxoids with at C-4(20) epoxide group are listed in Table 7. Most of these compounds carry Winterstein-type side chains at C-5, but an interesting group of three compounds from *Austrotaxus spicata* is esterified at C-9 with nicotinic acid.

3.3. Taxoids with an Oxetane Ring

The greatest interest from a medicinal point of view has been focused on the members of this class of taxoids which includes taxol and its potential semisynthetic precursors baccatin III and 10-deacetylbaccatin-III. Simple taxoids of this class are listed in Table 8. All members isolated to date are highly oxygenated, with oxygen substituents at C-2, C-4, C-7, C-9, C-10, and C-13. Even the C-1 position, which is normally not oxygenated in the other series of compounds, is usually oxygenated in this series.

Taxoids with both an oxetane ring and a complex C-13 ester side chain are listed in Table 9. This group, of course, includes the antitumor agents taxol and cephalomannine, but it also includes other congeners and a series of xylosyl derivatives isolated from *Taxus baccata* by POTIER and his collaborators (235). Not included in this Table are the many semisynthetic taxol side chain analogues that have been prepared; these compounds and their structure-activity relationships are however discussed in a recent review (131).

References, pp. 173–189

Table 6. *Neutral Taxoids with a C-4(20) Epoxide*

Name	Molecular formula	MW	R_1	R_2	R_3	R_4	R_5	R_6	$[\alpha]_D$	mp (°C)	Plant source	Plant part; % yield	References
Baccatin I (3.54)	$C_{32}H_{44}O_{13}$	636	H	Ac	OAc	H	Ac	Ac	+86°	298	T. baccata	l.s.r; 7.88×10^{-4}	(48)
5α-Deacetylbaccatin I (3.55)	$C_{30}H_{42}O_{12}$	594	H	H	OAc	H	Ac	Ac		256–258	T. baccata		(48)
1β-Hydroxybaccatin I (3.56)	$C_{32}H_{44}O_{14}$	652	OH	Ac	OAc	H	Ac	Ac	+102° +71.8° (CHCl₃)	257–262 273 (dec) 260–261	T. wallichiana T. baccata T. mairei	hw; 4.8×10^{-5}	(181) (48) (307)
1β-Hydroxy-5α-deacetylbaccatin I (3.57)	$C_{30}H_{42}O_{13}$	610	OH	H	OAc	H	Ac	Ac	+138.7° (CHCl₃)	273–275	T. yunnanensis	l.s; 4.3×10^4	(310)

Table 6. (Contd)

Name	Molecular formula	MW	R_1	R_2	R_3	R_4	R_5	R_6	$[\alpha]_D$	mp (°C)	Plant source	Plant part; % yield	References
1β-Acetoxy-5α-deacetylbaccatin I (3.58)	$C_{32}H_{44}O_{14}$	652	OAc	H	OAc	H	Ac	Ac		240–241	T. mairei	st b; 5.0×10^{-3}	(158)
1β-Hydroxy-7β-deacetoxy-7α hydroxy-baccatin I (3.59)	$C_{30}H_{42}O_{13}$	610	OH	Ac	H	OH	Ac	Ac	+74° (CHCl₃)	217–218	T. baccata	b; 2.4×10^{-3}	(235)
Spicatine (3.60)	$C_{35}H_{44}O_{10}$	624	H	a	H	H	H	Ac	+45° (CHCl₃)		Austrotaxus spicata	b; 3.9×10^{-4}	(69)
9α-Acetyl-10β-deacetyl-spicatine (3.61)	$C_{35}H_{44}O_{10}$	624	H	a	H	H	Ac	H	+17° (CHCl₃)		Austrotaxus spicata	b; 4.7×10^{-5}	(69)
10β-Deacetylspicatine (3.62)	$C_{33}H_{42}O_{9}$	582	H	a	H	H	H	H	+66° (CHCl₃)		Austrotaxus spicata	b; 7.8×10^{-5}	(69)

Table 7. *Basic Taxoids with a C-4(20) Epoxide*

Name	Molecular formula	MW	R₁	R₂	R₃	R₄	R₅	R₆	$[\alpha]_D$	mp (°C)	Plant source	Plant part; % yield	References
Spicataxine (3.63)	$C_{37}H_{51}NO_{10}$	669	a	H	Ac	OAc	CH_3	CH_3	$+67°$ $(CHCl_3)$		*Austrotaxus spicata*	b; 1.4×10^{-3}	(69)
9α-Acetyl 10β-deacetyl-spicataxine (3.64)	$C_{37}H_{51}NO_{10}$	669	a	Ac	H	OAc	CH_3	CH_3	$+43°$ $(CHCl_3)$		*Austrotaxus spicata*	b; 5.8×10^{-4}	(69)
Nicaustrine (3.65)	$C_{43}H_{54}N_2O_{11}$	774	a	b	Ac	OAc	CH_3	CH_3	$+67°$ $(CHCl_3)$		*Austrotaxus spicata*	b; 3.0×10^{-4}	(69)
N-Demethyl-nicaustrine (3.66)	$C_{42}H_{52}N_2O_{11}$	760	a	b	Ac	OAc	H	CH_3	$+93°$ $(CHCl_3)$		*Austrotaxus spicata*	b; 1.1×10^{-4}	(69)
Nicotaxine (3.67)	$C_{30}H_{37}NO_9$	555	H	b	Ac	=O	—	—	$+112°$ $(CHCl_3)$		*Austrotaxus spicata*	b; 6.3×10^{-5}	(69)
5α-O-(3'-methylamino 3'-phenylpropionyl) nicotaxine (3.68)	$C_{40}H_{48}N_2O_{10}$	716	a	b	Ac	=O	H	CH_3	$+103°$ $(CHCl_3)$		*Austrotaxus spicata*	b; 1.9×10^{-4}	(69)
5α-O-(3'-amino 3'-phenylpropionyl) nicotaxine (3.69)	$C_{39}H_{46}N_2O_{10}$	702	a	b	Ac	=O	H	H	$+93°$ $(CHCl_3)$		*Austrotaxus spicata*	b; 1.2×10^{-4}	(69)

Table 8. *Simple Taxoids with an Oxetane Ring*

Name	Molecular formula	MW	R_1	R_2	R_3	$[\alpha]_D$	mp (°C)	Plant source	Plant part; % yield	References
Baccatin IV (3.70)	$C_{32}H_{44}O_{14}$	652	OH	Ac	Ac	$+19°$	254–255 (dec)	*T. baccata*		(49)
1β-Acetylbaccatin IV (3.71)	$C_{34}H_{46}O_{15}$	694	OAc	Ac	Ac	$+59.3°$ (CHCl$_3$)	260–262	*T. yunnanensis*	l,s; 1.7×10^{-4}	(310)
1β-Dehydroxybaccatin IV (3.72)	$C_{32}H_{44}O_{13}$	636	H	Ac	Ac	$+5°$	286 (dec)	*T. baccata*		(49)
						$+99°$ (CHCl$_3$)	259–260	*T. mairei*	hw; 2.2×10^{-2}	(182)
1β-Dehydroxy-4α-deacetylbaccatin IV (3.73)	$C_{30}H_{42}O_{12}$	594	H	Ac	H	$+40$ (Acetone)	229–230	*T. mairei*	st.b; 1.7×10^{-3}	(158)
Baccatin VI (3.74)	$C_{37}H_{46}O_{14}$	714	OH	COC$_6$H$_5$	Ac	$-5°$ (CHCl$_3$)	248–250 (dec)	*T. baccata*	b; 4×10^{-4}	(235)
						$-9°$ (CHCl$_3$)	244–245 239–241	*T. baccata* *T. mairei*	hw; 3.2×10^{-3}	(49) (182)
1β-Dehydroxybaccatin VI (3.75)	$C_{37}H_{46}O_{13}$	698	H	COC$_6$H$_5$	Ac	$-21.2°$ (CHCl$_3$)	220–221	*T. mairei*	hw; 1.6×10^{-2}	(182)
Baccatin VII (3.76)	$C_{36}H_{52}O_{14}$	708	OH	COC$_5$H$_{11}$	Ac	$+9°$	270 (dec)	*T. baccata*		(49)

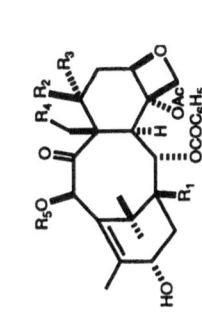

Name	Molecular formula	MW	R_1	R_2	R_3	R_4	R_5	$[\alpha]_D$	mp (°C)	Plant source	Plant Part; %yield	References
Baccatin III (3.77)	$C_{31}H_{37}O_{11}$	585	OH	OH	H	H	Ac	$-54°$ (CH₃OH) / $-54°$ (CH₃OH)	236–238 (dec) / 229–231	T. baccata / T. wallichiana	b; 3.2×10^{-4}	(235) / (181, 213)
10-Deacetylbaccatin III (3.78)	$C_{29}H_{35}O_{10}$	543	OH	OH	H	H	H	$-41°$	242–245 / 221–223	T. baccata / T. yunnanensis	l; 2.0×10^{-2} / l,s; 1.4×10^{-3}	(35) / (310)
								$-249°$ (EtOAc)	223–225	T. brevifolia	b; 1.0×10^{-4}	(132)
19-Hydroxybaccatin III (3.79)	$C_{31}H_{37}O_{12}$	601	OH	OH	H	OH	Ac		180–182 / 173–175	T. baccata / T. yunnanensis	l; 2.0×10^{-3} / l,s; 1.7×10^{-3}	(35) / (310)
									171–174	T. wallichiana	r,s,l; 9.1×10^{-4}	(175)
1-Dehydroxybaccatin III (3.80)	$C_{31}H_{37}O_{10}$	569	H	OH	H	H	Ac	$-45.9°$ (CHCl₃)	169–171	T. yunnanensis	l,s; 2.7×10^{-4}	(310)
Baccatin V (3.81)	$C_{31}H_{37}O_{11}$	585	OH	H	OH	H	Ac	$-87°$	254–255	T. baccata		(47)

Table 9. *Taxoids with an Oxetane Ring and a complex C-13 Side Chain*

Name	Molecular formula	MW	R_1	R_2	R_3	R_4	$[\alpha]_D$	mp (°C)	Plant source	Plant part; % yield	References
Taxol (3.82)	$C_{47}H_{51}NO_{14}$	853	OH	H	Ac	C_6H_5	−49° (CH₃OH)	213–216 (dec)	*T. brevifolia*	st b; 2×10^{-2}	(285)
							−54° (CH₃OH)	194–197	*T. brevifolia*	b; 2.9×10^{-4}	(132)
							−42° (CH₃OH)	198–203	*T. wallichiana*	s,r	(213)
							−42° (CH₃OH)	198–203	*T. wallichiana*	l,s,r; 1.1×10^{-3}	(181)
							−54° (CH₃OH)	205–208	*T. baccata*	b; 1.65×10^{-2}	(235)
							−21° (py)				
7-*epi*Taxol (3.83)	$C_{47}H_{51}NO_{14}$	853	H	OH	Ac	C_6H_5	−32.3° (CH₃OH)	168–171	*T. brevifolia*	b; 1.79×10^{-3}	(113)
10-Deacetyltaxol (3.84)	$C_{45}H_{49}NO_{13}$	811	OH	H	H	C_6H_5			*T. wallichiana*	r,s,l; 2.1×10^{-3}	(175)
							−3° (py)		*T. baccata*	b; 2.9×10^{-3}	(235)
7-(β-Xylosyl)-taxol (3.85)	$C_{52}H_{59}NO_{18}$	985	a	H	Ac	C_6H_5	−23° (py)	236–238	*T. baccata*	b; 5.8×10^{-3}	(235)
7-(β-Xylosyl)-10-deacetyltaxol (3.86)	$C_{50}H_{57}NO_{17}$	943	a	H	H	C_6H_5	−2° (py)	246–248	*T. baccata*	b; 2.2×10^{-2}	(235)

Compound	Formula	MW				Side chain	$[\alpha]$	mp	Source	Activity	Ref.
10-(β-hydroxybutyryl)-10-deacetyltaxol (**3.87**)	$C_{49}H_{55}NO_{15}$	897	OH	H	b	C_6H_5			*T. baccata*	b; 2.6 × 10⁻³	(235)
Cephalomannine (**3.88**)	$C_{45}H_{53}NO_{14}$	831	OH	H	Ac	[structure]	−41° (CH₃OH)	181–184	*T. baccata*	b; 6.4 × 10⁻³	(235)
							−41° (CH₃OH)	184–186	*T. wallichiana*	l,s,r; 1.63 × 10⁻³	(181, 213)
								180–183	*T. baccata*	b; 3 × 10⁻³	(35)
10-Deacetylcephalomannine (**3.89**)	$C_{43}H_{51}NO_{13}$	789	OH	H	H	[structure]			*T. wallichiana*	r,s,l; 3.8 × 10⁻³	(175)
							−2°(py)		*T. baccata*	b; 3.4 × 10⁻³	(235)
7-(β-Xylosyl)-cephalomannine (**3.90**)	$C_{50}H_{61}NO_{18}$	963	a	H	Ac	[structure]	−26°(py)		*T. baccata*	b; 3.9 × 10⁻³	(235)
7-(β-Xylosyl)-10-deacetyl-cephalomannine (**3.91**)	$C_{48}H_{59}NO_{17}$	921	a	H	H	[structure]	+4°(py)	250–252	*T. baccata*	b; 9.6 × 10⁻³	(235)
10-(β-hydroxybutyryl)-10-deacetylcephalomannine (**3.92**)	$C_{47}H_{57}NO_{15}$	875	OH	H	b	[structure]			*T. baccata*	b; 1.8 × 10⁻³	(235)
7-(β-Xylosyl)-taxol C (**3.93**)	$C_{51}H_{65}NO_{18}$	979	a	H	Ac	C_5H_{11}	−4°(py)	229–231	*T. baccata*	b; 2.9 × 10⁻³	(235)
7-(β-Xylosyl)-10-deacetyltaxol C (**3.94**)	$C_{49}H_{63}NO_{17}$	937	a	H	H	C_5H_{11}	+3°(py)	215–217	*T. baccata*	b; 2.8 × 10⁻³	(235)

Table 10. *Miscellaneous Taxoids*

Name	Molecular formula	MW	R	[α]$_D$	mp (°C)	Plant source	Plant part; % yield	References
Taxagifine III (3.95)	$C_{24}H_{34}O_{11}$	498	Ac	+31.4° (CH$_3$OH)	246–247	T. chinensis	l,s; 1.5 × 10^{-4}	(311, 312)
4-Deacetyltaxagifine III (3.96)	$C_{22}H_{32}O_{10}$	456	H	+38.1° (CH$_3$OH)	221–223	T. chinensis	l,s; 2.0 × 10^{-4}	(311)

Name	Molecular formula	MW	R	[α]$_D$	mp (°C)	Plant source	Plant part; % yield	References
Taxinine K (3.97)	$C_{26}H_{36}O_8$	476	H		167–168	T. cuspidata	l; 1.6 × 10^{-5}	(39)
Taxinine L (3.98)	$C_{28}H_{38}O_9$	518	Ac		159–160	T. cuspidata	l; 1.2 × 10^{-5}	(39)

Name		MW	Molecular formula	[α]$_D$	mp (°C)	Plant source	Plant part; % yield	References
Spicaledonine (3.99)		667	$C_{37}H_{49}NO_{10}$	+ 29° (CHCl$_3$)	a	Austrotaxus spicata	b; 9.3 × 10^{-5}	(69)

Name	MW	Molecular formula	[α]$_D$	mp (°C)	Plant source	Plant part; % yield	References
10-Deacetyl-10-oxo-7-epi-taxol (3.100)	809	$C_{45}H_{47}NO_{13}$	− 60.4° (CH$_3$OH)		T. brevifolia	b; 1.07 × 10^{-3}	(113)

Name	MW	Molecular formula	[α]$_D$	mp (°C)	Plant source	Plant part; % yield	References
Taxine A (3.101)	641	$C_{35}H_{47}NO_{10}$	− 140° (CHCl$_3$)	204–206	T. baccata	1	(82, 83)

3.4. Miscellaneous Taxoids

A diverse group of miscellaneous taxoids is listed in Table 10. Of particular interest is a group of three compounds with a C-3(11) bond. Compounds of this type can be prepared by irradiation of taxinine derivatives (*39, 143*) but they have also been isolated as natural products. It is certainly possible that the naturally occurring substances are formed in the plant by irradiation or by sunlight, but the details of their formation have not been established.

Also of interest is the alkaloid taxine A, which does not have the normal taxane skeleton.

4. The Chemistry of the Taxoids

4.1. Isolation Techniques

4.1.1. Extraction

As noted earlier, the original workers in the field were interested primarily in the toxic alkaloidal constituents of the yew and hence used extraction procedures appropriate for basic compounds. Thus LYTHGOE, in a procedure typical of those used by earlier workers, soaked yew needles seven days in 0.65% H_2SO_4 (*5*) in order to obtain the basic Winterstein esters which he then converted to O-cinnamoyltaxicin-I triacetate. These relatively harsh conditions probably result in hydrolysis or other changes in some of the more labile taxoids present; thus baccatin III (**3.77**) is converted to baccatin V (**3.81**) under mild conditions (*213*) and similar conversions could occur with other taxoids under the strongly acidic conditions.

Most of the recent work on the isolation of taxoids has been carried out by means of extraction using neutral solvents. Petroleum ether extraction of the heartwood of *T. baccata* yielded the baccatins (*269*), and alcohol was used to extract taxol from *T. brevifolia* (*285*). Other investigators have used extraction solvents somewhere between these two polarity extremes. The importance of taxol has meant that its extraction has been given particular attention, but unfortunately most of this work is proprietary and has not been published. However, a recent paper (*121*) describes a supercritical fluid extraction of taxol from the bark of *T. brevifolia*. The use of carbon dioxide and ethanol mixtures at 318K gave a recovery of between 50 and 85% of the total taxol present, and furnished an extract in which the taxol represented up to 1.5% of the total extract, as compared with 0.125% for an ethanol extract. The use of supercritical fluid extraction is thus a promising technique for selective extraction of taxol from plant material.

4.1.2. Purification and Analysis

Purification of the taxoids has relied primarily on chromatographic techniques. The details of the individual isolations are contained in the references provided in the preceding tabular survey of taxane structures and will not be repeated here.

The isolation of taxol itself has obviously been studied very thoroughly, but here again much of this work is proprietary and unpublished. One current procedure uses sequential chromatography on silica gel and Florisil, yielding a mixture of taxol and cephalomannine which is finally separated by careful chromatography on silica gel using dichloromethane/1-propanol as eluant. In our hands this procedure yields pure taxol in the first fractions, but later fractions consist of a mixture of taxol and cephalomannine. A clean separation of taxol from cephalomannine can, however, be accomplished by converting the cephalomannine (**4.1**) in the mixture to its diol derivative (**4.2**) (Scheme 4.1) (*133*); taxol is not affected by these conditions and can be separated from the diol by simple flash chromatography. An alternate separation of taxol and cephalomannine can be accomplished by HPLC on a cyanopropyl column with hexane: isopropanol (2:1) (*29*). Recently a preparative separation of taxol, cephalomannine, and baccatin III has been reported using high-speed countercurrent chromatography and a solvent system of hexane:ethyl acetate:methanol:water, 1:1:1:1 (*215*).

Analytical separations of taxol and related compounds of importance such as baccatin III, 10-deacetylbaccatin-III, and cephalomannine have been developed by several groups. Most of these have involved reverse-phase HPLC, using either cyano or phenyl bonded phase columns (*301*)

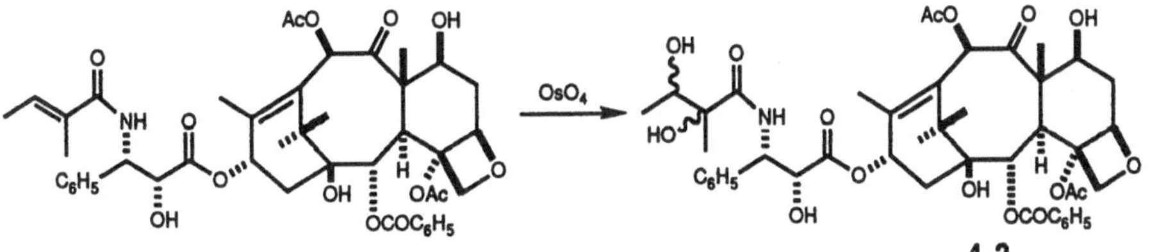

4.1   4.2

Scheme 4.1. Conversion of cephalomannine to its diol derivative

or C_{18} bonded phase columns (283). Thus the first group selected a phenyl column with a linear gradient from $MeOH:H_2O$: $MeCN(20:65:15)$ to $MeOH:H_2O:MeCN$ (20:45:35) and ending with $MeOH: H_2O:MeCN$ (20:25:55), while the second group used a C_{18} column with $MeOH: H_2O$ (68:32) as the solvent. Recently a separation on a perfluorophenyl column has been reported (179), and a method for analysis of taxol in *T. chinensis* has also been described (303).

A novel approach to the analysis of taxoids in plant material has been developed by COOKS and his collaborators (104). These workers have used MS/MS methods and multiple reaction monitoring parent scans to detect taxol, cephalomannine, and baccatin III with limits of 100 pg and linear calibration curves.

Analysis for taxol in biological fluids has also made use of HPLC. Analysis of taxol in human plasma and urine was carried out on a C_8 column with MeOH:NaOAc buffer (0.02M, pH 4.5, 35:65 v/v) (218), or on a C_{18} column with a gradient of H_2O: MeCN (65:35 to 0:100) (162). A reverse-phase HPLC method was also used for analysis of the decomposition products of taxol in cell culture media (216) and separation of metabolites in rat bile (184).

The presence of taxol in plant extracts can also be detected by two-dimensional TLC methods (251). One system uses a diphenyl modified silica gel plate, with elution by hexane : iPrOH: CH_3COCH_3 (15:2:3) in one direction and $MeOH: H_2O$ (7:3) in the other. The other system uses a cyano modified silica gel plate, with CH_2Cl_2: hexane: AcOH (9:10:1) in one direction and H_2O: MeCN: MeOH: THF (8:5:7:0.1) in the other.

Initial results on an immunoassay for taxol have been published (120), but full details of this method are not yet available.

Studies of the content of both taxol and the related diterpenoids baccatin III (**1.10**) and 10-deacetylbaccatin III (**4.3**) have been reported. The latter two diterpenoids are important potential precursors to taxol, since both can be converted to taxol by methods that will be described later. The amounts of taxol and 10-deacetylbaccatin III in six *Taxus* species were analyzed by HPLC (301). The highest taxol content of 0.01% was found in needles of *T. X media* cv. Hicksii, but comparable

4.3

amounts were found in needles of several other species. The highest yield of 10-deacetylbaccatin III (0.02%) was found in *T. baccata* cv. repandens needles.

A second study surveyed more plant samples, but reported only on taxol content (*283*). Taxol content was found to be highly variable, with 15 bark samples of *T. brevifolia* giving taxol contents between 0.001 and 0.069% and the needles giving taxol contents of 0.0003–0.003%. Although taxol content was highest in *T. brevifolia* bark, *T. brevifolia* seedlings and *T. media* twigs also contained respectable amounts.

The third study used the ELISA technique to make semiquantitative determinations of taxol-like materials in various *Taxus* species. Both leaves and stem bark of *T.baccata* gave results indicating yields up to 0.05% for leaves and 0.08% for stem bark (*120*).

Two additional studies of taxol content have appeared recently. A study of *Taxus cuspidata* (*191*) showed that taxol was concentrated in the bark (0.012% of dry weight) and needles (0.035% dry weight). The effect of different needle post-harvesting procedures was also evaluated; storage for 4 days at 60 °C led to some loss of taxol. A study of *Taxus brevifolia* (*291*) showed that significant variation in taxol content exists among and within populations and species, but that taxol contents exceeding those reported for *T. brevifolia* bark can be found in shoots of individual trees. The season of collection and sample handling techniques also affect taxoid content.

The yield of 10-deacetylbaccatin III obtainable from *Taxus* species is probably much higher than indicated from the data cited above. Thus it was isolated from *T. baccata* trunk bark in a yield of 0.02% (*35*) and it has been isolated from European *T. baccata* needles in yields of 0.1% (*55*), although no experimental details are provided. It thus appears likely that selection of suitable plant material and appropriate genetic selection techniques will lead to the availability of 10-deacetylbaccatin III in yields of at least 0.1% from *Taxus* needles, and even higher yields are probable.

4.2. Spectroscopy

4.2.1. UV and ORD/CD Spectroscopy

The UV spectra of taxoids containing only a ring A enone system display an anomalously long wave length absorption band. Thus 5-deoxy-4, 16-dihydrotaxicin-l (**4.4**) has λ_{max} (EtOH) 283 nm (ε 5700), while the expected value for an enone such as this would be about 255 nm (*97, 164*). This bathochromic shift appears to be due to a combination of factors. One factor is the presence of the C-1 hydroxyl group, since the corresponding taxicin-II derivatives (which lack this hydroxyl group) absorb at about 7 nm shorter wavelength. The overall ring strain of the taxane skeleton is, however, the major factor and this factor is apparent in the spectrum of the alcohol (**4.5**) which absorbs at 227 nm as opposed to 202 nm predicted for an unstrained alkene. Opening of ring C does not make a major difference in the spectrum and the ring-opened product (**4.6**) has λ_{max} 276 nm. This is consistent with MM2 results mentioned previously (*257*), since the A/B ring system alone accounts for the bulk of the strain energy of the taxene ring. Ring-strain alone cannot account for the whole effect, however, since ketone (**4.7**) has a normal λ_{max} of 250 nm (*214*), and it has been suggested that the *gem*-dimethyl group at C-15 is implicated (*164*). The optical rotary dispersion (ORD) and circular dichroism (CD) spectra of the taxane diterpenoids have provided valuable confirmatory evidence for their stereochemistry. As mentioned earlier, initial stereochemical assignments of taxinine (**1.3**) were made on the basis of ^{1}H-NMR and ORD data (*152*). Application of the benzoate

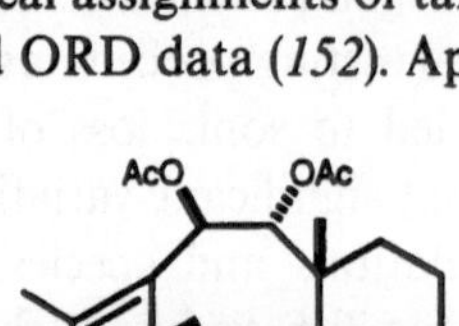

4.4

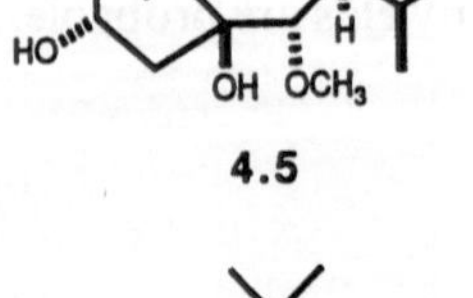

4.5 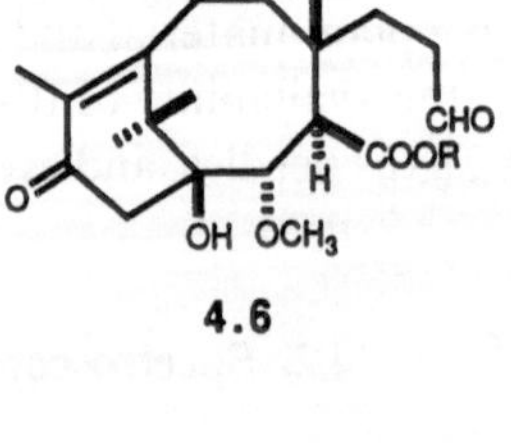   **4.6** 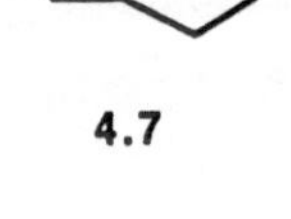   **4.7**

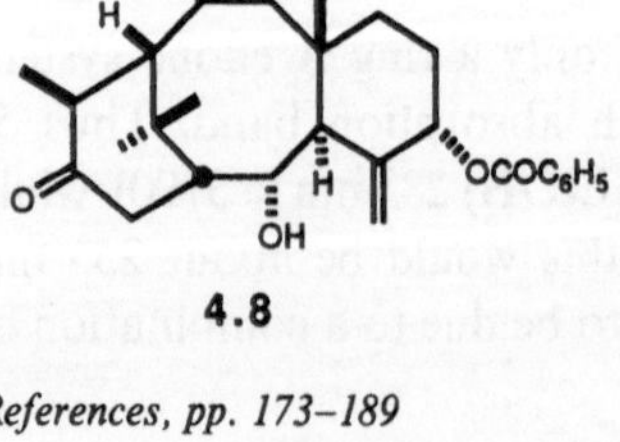

4.8 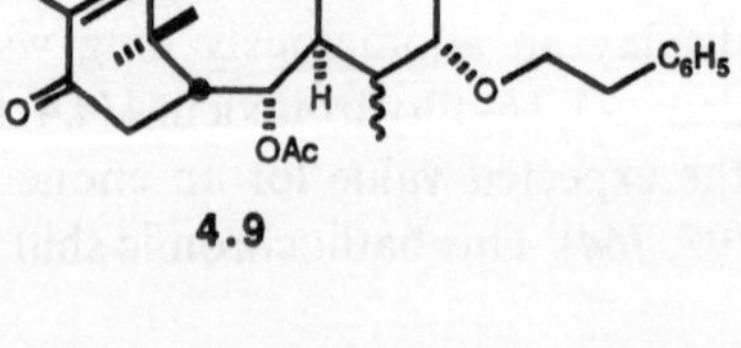   **4.9**

sector rule to 5-benzoyl-isopropylidenetaxinol (**4.8**) confirmed its stereochemistry (*95*), and the application of the rather more powerful dibenzoate chirality rule to the tetrahydrotaxinine dibenzoate (**4.9**) confirmed its
stereochemistry (*94*). Thus the dibenzoate (**4.9**) showed a negative first
Cotton effect at 234 nm, corresponding to a negative benzoate chirality
and thus to the stereochemistry shown. An independent investigation of
the Cotton effects associated with the bridgehead olefin of several taxoids
yielded similar conclusions (*51*). This study, which was carried out on
taxoids lacking a C-13 carbonyl group, showed that these compounds all
exhibited intense positive Cotton effects in the 214–230 nm region due
primarily to the $\pi \to \pi^*$ olefinic absorption. These effects corresponded
to those observed with (S)-(+)-*trans*-cyclodecene, in accordance with
the previous stereochemical results.

4.2.2. Infrared Spectroscopy

The taxoids do not show any unusual infrared absorption properties.
Thus a simple taxicin derivative such as 5-deoxy-4, 20-dihydrotaxicin-1
(**4.10**) shows a normal IR-absorption at 1664 (conjugated ketone) and
1597 cm^{-1} (double bond) (*164*).

4.10

4.2.3. ^{1}H-NMR Spectroscopy

Studies of the ^{1}H-NMR spectra of the taxoids through 1979 have
been well reviewed by MILLER (*180*), and this material will thus not be
reviewed again. Interest over the last decade has centered on the
^{1}H-NMR spectrum of taxol which is reproduced in Fig. 2. Assignments
for the proton signals in taxol have been reported by several investigators
(*132, 180, 181*), but recent detailed studies at 500 MHz (*40, 73*) provide
the most complete analysis of the spectrum. The assignments made in
these studies were confirmed by NOE measurements, and are largely
consistent with previous assignments made at lower field strengths,

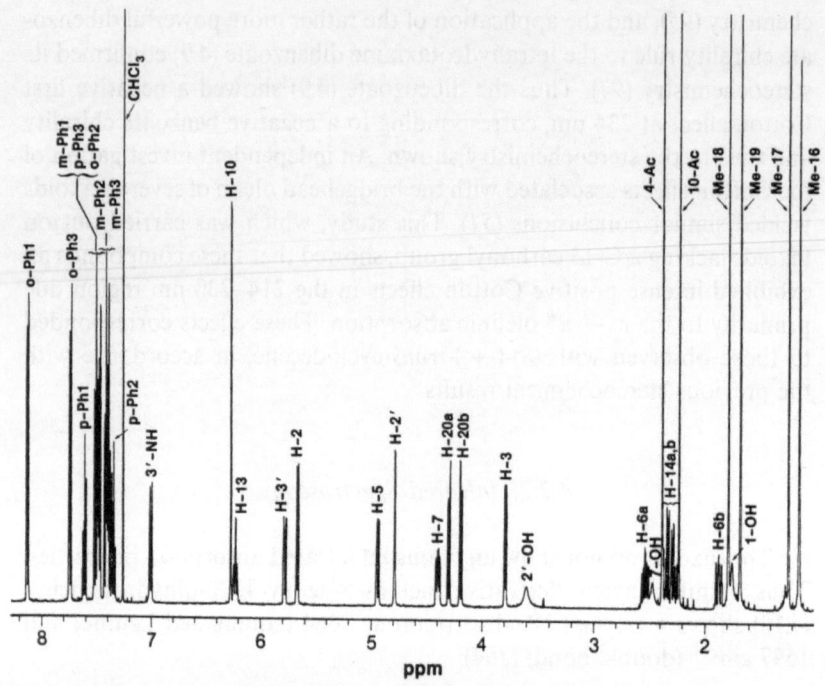

Fig 2. 500 MHz ¹H-NMR spectrum of taxol

except that the assignment of the 16- and 17-methyl groups are reversed by one study (40). The complete assignments of this group are shown in Fig. 3. One important point about the ¹H-NMR spectrum of taxol is that it is quite solvent dependent. Thus in CDCl₃ the C-20 protons appear as a well-resolved AB quartet with some additional long-range coupling to H-2 and H-3 detectable (40). However, these signals appear as a broad singlet in CD₃OD (166). ¹H-NMR spectra of selected taxoids are presented in Tables 11–18. The data in these Tables are taken directly from the cited literature references, and the assignments have thus been made with varying degrees of certainty.

The ¹H-NMR spectrum of taxol has also been obtained by a novel solid/liquid intermolecular transfer dynamic nuclear polarization technique (59). In this method, dynamic nuclear polarization enhancement was generated at low field using nitroxides immobilized on silica gel in contact with the dissolved sample. The polarized sample was then monitored with improved spectral resolution at high field. Under the experimental conditions used, ¹H-NMR signals with long spin-lattice

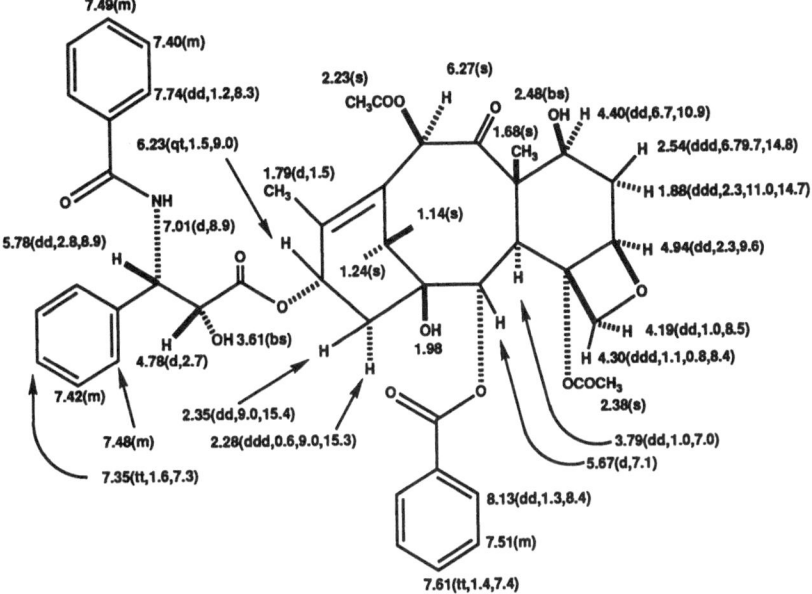

Fig. 3. ¹H-NMR assignments for taxol

relaxation times, such as the aromatic protons, showed dipolar domi-
nated or emission enhancements, while signals with short relaxation
times showed only small enhancements. In the experiment described, the
C-10 acetate and the C-16, C-17, and C-19 methyl groups showed larger
enhancements than the C-4 acetate and the C-18 methyl group, sugges-
ting that the upper or convex face of the taxol molecule is the one that
interacts most strongly with the silica surface.

4.2.4. ¹³C-NMR Spectroscopy

Assignments have been made of the ¹³C-NMR spectra of various
$\Delta^{4(20),11}$ taxadiene derivatives (52) and of various oxirane and oxetane-
containing taxoids (219). These assignments, together with assignments
of the ¹³C-NMR spectrum of taxol (166), were made by conventional 1D
methods. A recent paper by BEUTLER (40) reports the assignment of the
¹³C-NMR spectrum of taxol by 2D-methods; these assignments are
shown in Fig. 4. The ¹³C-NMR spectra of selected taxoids are given in
Tables 19–20.

Table 11. ¹H NMR Spectra of Selected Taxoids ᵃ· ᵇ

Protons on	Taxinine J (3.37) (182)	5α-Cinnamoyloxy-2α,13α-dihydroxy-9α,10β-diacetoxy-4(20),11-taxadiene (3.31) (311)	Decinnamoyltaxinine J (3.8) (132)	7β,9α,10β-Triacetoxy-2α,5α,13α-trihydroxy-4(20),11-taxadiene (3.5) (69)	7β,9α-Diacetoxy-5α,13α,14β-trihydroxy 10-oxo 4(20),11-taxadiene (3.46) (69)	Brevifoliol (3.16) (2)
C-1	1.90 (br d, 9.2)	1.94	1.94	1.96 (br d, 9)		1.50 (d)
C-2	5.55 (br d, 7.0)	4.13 (dd, 6, 1)	5.46 (dd, 6, 2)	3.16 (d, 5)		2.36 (dd, 14.1, 9.2)
C-3	3.35 (d, 7.0)	3.19 (d, 6)	3.49 (d, 6)	4.20 (d, 5)	3.02 (br s)	2.77 (br d)
C-5	5.47 (m)	5.83 (t, 5)	4.24 (br t)	4.32 (s)	4.35 (m)	4.43 (br s)
C-6α						1.80 (m)
C-6β		2.00 (m)	2.0 (m)	1.67 (m)	1.57 (br t, 11)	2.00
C-7	5.54 (dd, 10, 6.5)	1.35 (m)	5.51 (dd, 10, 4)	5.51 (dd, 11, 5)	5.13 (dd, 11, 5)	5.57 (dd, 11.3, 5.1)
C-9	5.90 (d, 10.4)	5.45 (d, 10)	5.84 (d, 9)	5.72 (d, 11)	5.56 (s)	6.05 (d, 10.5)
C-10	6.22 (d, 10.4)	5.86 (d, 10)	6.21 (d, 9)	6.10 (d, 11)		6.53 (d, 10.5)
C-13	5.85 (br t, 7.0)	4.25 (br d, 10)	5.71 (tq, 4, 1)	4.34 (br s)	4.35 (m)	4.38 (br t, 7.2)
C-14α	1.50 (br dd, 15.2, 8.0)	1.93 (m)	2.6 (m)	1.41 (br d, 16)	3.65 (br s)	1.30 (dd, 13.9, 7.2)
C-14β	2.70 (ddd, 15.2, 9.2, 8.0)	2.64 (ddd, 15, 10, 9)		2.73 (dt , 16, 10)		2.46 (dd, 13.9, 7.3)
C-16-Me	1.76 (s)	1.16 (s)	1.72 (s)	1.60 (s)	1.46 (s)	1.05 (s)
C-17-Me	1.12 (s)	1.14 (s)	1.03 (s)	0.99 (s)	1.08 (s)	1.35 (s)

C-18-Me	2.33 (s)		2.23 (d, 1)		2.27 (s)	2.06 (s)	1.74 (s)
C-19-Me	1.03 (s)		0.96 (s)		0.94 (s)	0.90 (s)	0.90 (s)
C-20	5.00 (br s) 5.48 (br s)		4.84 (br s) 5.28 (br s)		5.49 (s) 5.31 (s)	4.88 (s) 5.21 (s)	4.82 (br s) 5.18 (br s)
C-2'	6.60 (d, 16.0)		6.67 (d, 16)				
C-3'	7.80 (d, 16.0)		7.79 (d, 16)				
Ph	7.40 (m) 7.50 (m)						7.43 7.55 7.87
OAc	2.03 (s) 2.07 (s) 2.07 (s)	2.10 (s)	1.98 (s) 2.03 (s) 2.05 (s)	2.05 (s) 2.09 (s)	1.99 (s) 2.03 (s) 2.06 (s)	2.09 (s) 2.17 (s)	2.01 (s) 2.07 (s)

[a] Measured in CDCl$_3$. Chemical shifts (γ) are expressed in parts per million from Me$_4$Si and coupling (J) in hertz. [b] Multiplicity: s = singlet, d = doublet, t = triplet, q = quartet, m = multiplet, br = broad.

Table 12. ^1H NMR Spectra of Selected Taxoids[a,b]

Protons on	Taxine B (3.29) (70)	Austrospicatine (3.19) (68)	2'β-Deacetoxyaustro-spicatine (3.24) (68)	Austrotaxine (3.42) (68)	Comptonine (3.28) (69)
C-1			1.84 (m)		
C-2α	3.98 (d, 7)	1.88 (dd, 16, 6)	1.84 (dd)		
C-2β			1.69 (dd)		
C-3	3.17 (d, 7)	2.78 (br d, 6)	2.79 (br d, 5)	2.7 (br d, 6)	1.95 (m)
C-5	5.10 (br s)	5.68 (dd, 2, 1)	5.36 (m)	5.76 (m)	2.71 (br d, 4)
C-6α	1.57 (m)	1.79 (m), 1.68 (m)	1.25 (dd)	1.76 (dd, 8, 3)	5.10 (br s)
C-6β			1.60 (m)		1.52 (m)
C-7	1.78 (m), 1.37 (m)	5.59 (dd, 11, 5)	5.40 (dd, 11.5, 5)	5.28 (t, 8)	5.29 (dd, 11, 5)
C-9	4.28 (d, 9.5)	6.00 (d, 11)	5.87 (d, 11.5)	5.76 (m)	5.85 (d, 11)
C-10	5.85 (d, 9.5)	6.28 (d, 11)	6.28 (d, 11.5)		6.21 (d, 11)
C-13		5.89 (br t, 8)	5.87 (br t)	6.31 (br d, 7)	
C-14α	2.65 (s)	2.60 (dt, 14, 10)	0.96 (dd, 14.5, 7.5)		1.95 (m)
C-14β			2.65 (dt, 14.5, 9.5)	4.66 (d, 7)	2.96 (dd, 20, 7)
C-16-Me	1.52 (s)	1.64 (s)	1.60 (s)	1.55 (s)	1.62 (s)
C-17-Me	1.25 (s)	1.13 (s)	1.10 (s)	1.35 (s)	1.14 (s)
C-18-Me	2.15 (s)	2.25 (s)	2.20 (d, 1)	2.0 (br s)	2.42 (s)
C-19-Me	1.13 (s)	0.85 (s)	0.77 (s)	0.98 (s)	0.78 (s)

C-20	5.40 (s) 5.35 (s)	5.36 (s) 5.01 (s)	4.94 (br s) 5.25 (d, 1)	5.53 (s) 5.35 (s)	5.24 (s) 4.92 (s)
C-2'	2.92 (dd, 16, 7) 2.53 (dd, 16, 7)	5.41 (d, 4)	2.92 (m)	5.45 (d, 4)	4.55 (d, 9)
C-3'	3.85 (t, 7)	4.21 (d, 4)	4.0 (m)	4.16 (d, 4)	3.96 (m)
(Me)$_2$N	2.22 (s)	2.27 (s)	2.25 (m)	2.21 (s)	2.51 (s)
3'-Ph (ortho)	7.32 (m)	7.37 (m)	7.32 (m)	7.36 (m)	7.36 (m)
3'-Ph (meta)					
3'-Ph (para)			7.24 (m)		
2-OAc		1.41 (s)		1.48 (s)	
OAc	2.22 (s)	2.00 (s) 2.04 (s) 2.11 (s)	2.05 (s) 2.01 (s) 2.01 (s) 1.97 (s)	2.03 (s) 2.15 (s) 2.20 (s)	2.05 (s) 2.06 (s) 2.07 (s)

[a] Measured in $CDCl_3$. Chemical shifts (δ) are expressed in parts per million from Me_4Si and coupling constants (J) in hertz. [b] Multiplicity: s = singlet, d = doublet, t = triplet, q = quartet, m = multiplet, br = broad.

Table 13. ^1H NMR Spectra of Selected Taxoids[a, b]

Protons on	Spicataxine (3.63) (69)	Spicatine (3.60) (69)	Nicaustrine (3.65) (69)	Nicotaxine (3.67) (69)
C-1	1.54 (m)	1.69 (m)	1.70 (d, 9)	2.09 (br d, 7)
C-2	5.30 (br d, 3.5)	5.44 (br d, 3)	5.51 (br d, 3.8)	5.65 (d, 2.3)
C-3	2.84 (br d, 3.5)	3.07 (d, 3)	3.0 (d, 3.8)	3.23 (d, 3)
C-5	4.13 (br s)	4.23 (br s)	4.20 (br s)	2.98 (d, 2)
C-6α	1.67 (m), 1.91 (m)	C		
C-6β		C		
C-7	1.47 (m)			
C-9	4.26 (d, 10)	4.36 (d, 10)	6.17 (d, 11)	6.20 (d, 10)
C-10	5.73 (d, 10)	5.85 (d, 10)	6.20 (d, 11)	6.28 (d, 10)
C-13	5.75 (br t, 9.7)	6.03 (br t, 8)	5.81 (t, 8)	
C-14α	1.07 (m)	1.69 (m)	1.17 (dd, 16, 8)	2.45 (d, 20)
C-14β	2.38 (dt, 15.5, 9.5)	2.77 (dt, 15, 10)	2.51 (dt, 16, 9)	2.93 (dd, 20, 7)
C-16-Me	1.52 (s)	1.63 (s)	1.80 (s)	1.81 (s)
C-17-Me	1.01 (s)	1.16 (s)	1.12 (s)	1.18 (s)
C-18-Me	2.11 (br s)	2.23 (s)	2.23 (s)	2.28 (s)
C-19-Me	1.24 (s)	1.34 (s)	1.22 (s)	1.18 (s)
C-20	2.24 (d, 5.4)	2.36 (d, 5)	2.32 (d, 5)	2.18 (d, 5)
	3.33 (d, 5.4)	3.52 (d, 5)	3.40 (d, 5)	3.56 (d, 5)

C-2'	2.72 (dd, 14.8, 8.3) 3.05 (dd, 14.8, 6.5)	7.76 (d, 16)	2.73 (dd, 14.8, 7.9) 3.09 (dd, 14.8, 6.8)	
C-3'	3.89 (dd, 8.3, 6.5)	6.45 (d, 16)	3.92 (dd, 7.9, 6.8)	
C-3'-Ph	7.28–7.30 (m)	7.43–7.51 (m)	7.28–7.33 (m)	
Me$_2$N	2.20 (s)		2.24 (s)	2.04 (s)
OAc	2.20 (s) 2.06 (s) 1.93 (s)	2.23 (s) 2.16 (s) 2.01 (s)	2.28 (s) 2.01 (s) 1.80 (s)	1.84 (s)
C-9 Nicotinoyl				
2'''			9.19 (dd, 2, 0.7)	9.26 (br s)
4'''			8.26 (ddd, 7.9, 2, 1.7)	8.31 (br d, 8)
5'''			7.40 (ddd, 7.9, 4.8, 0.7)	7.43 (dd, 8, 4)
6'''			8.78 (dd, 4.8, 1.7)	8.84 (br d, 4)

[a] Measured in CDCl$_3$. Chemical shifts (δ) are expressed in parts per million from Me$_4$Si and coupling constants (J) in hertz.
[b] Multiplicity: s = singlet, d = doublet, t = triplet, q = quartet, m = multiplet, br = broad. [c] Signals not identified.

D.G.I. KINGSTON, A.A. MOLINERO and J.M. RIMOLDI

Table 14. ¹H NMR Spectra of Selected Taxoids[a,b]

Protons cn	Taxagifine[c] (3.48) (34)	2α-Deacetyl-5α-decinnamoyltaxagifine (3.49) (311)	5α-Acetyl-5α-decinnamoyl-taxagifine (3.51) (314)	Taxinine M (3.52) (15)
C-1	2.34 (dd, 11, 6)	2.77 (dd, 11, 1)	2.34 (dd, 11.6, 1.5)	2.48 (ddd, 11.6, 2.4, 0.7)
C-2	5.40 (m)	4.23 (dd, 9, 1)	5.52 (dd, 9, 1.5)	6.14 (dd, 10.4, 2.4)
C-3	3.36	3.34 (d, 9)	3.30 (d, 9)	3.71 (d, 10.4)
C-5	5.40 (m)	4.43 (t, 3)	5.25 (t, 3, 2.5)	4.45 (br t, 2–3)
C-6α	2.20 (m)	2.15 (m)	2.18 (ddd, 14, 6, 2.5)	2.23 (ddd, 14.2, 6.1, 2.1)
C-6β	1.64 (d, 14)	1.59 (ddd, 14.5, 10.5, 3)	1.56 (ddd, 14, 11, 3)	1.70 (ddd, 14.2, 10.7, 3.6)
C-7	5.40 (m)	5.29 (dd, 10.5, 6)	5.30 (dd, 11, 6)	5.51 (dd, 10.7, 6.2)
C-9	4.92 (d, 2.5)	4.93 (d, 3)	4.88 (d, 3)	5.36 (d, 3)
C-10	5.36 (d, 2.5)	5.26 (d, 3)	5.35 (d, 3)	5.31 (d, 3)
C-11 OH				4.10 (br s)
C-14α	2.5 (d, 18)	2.73 (d, 18.4)	2.50 (d, 18.3)	3.00 (dd, 19.2, 11.6)
C-14β	3.0 (d, 18)	3.06 (dd, 18.4, 11)	3.07 (dd, 18.3, 11.6)	2.75 (d, 19.2)
C-16	3.66 (d, 8)	3.78 (d, 8)	3.69 (d, 8)	3.63 (d, 8.1)
	4.16 (d, 8)	4.17 (d, 8)	4.19 (d, 8)	4.08 (d, 8.1)
C-17-Me	1.52 (s)	1.47 (s)	1.52 (s)	1.29 (s)
C-18-Me	1.08 (s)	1.13 (s)	1.17 (s)	1.17 (s)
C-19-Me	1.20 (s)	1.05 (s)	1.09 (s)	5.14 (d, 12.2)
				4.42 (d, 12.2)

C-20	4.56 (s) 5.40 (s)	5.17 (s) 5.50 (s)	4.60 (s) 5.05 (s)	4.68 (s) 5.41 (s)
C19-OBz (ortho) (meta) (para)				8.16 7.51 7.61
C-2'	6.88 (d, 16)			
C-3'	7.90 (d, 16)			
3'-Ph	} 7.26–7.90 (m)			
OAc	1.97 (s) 2.09 (s) 2.11 (s)		1.96 (s) 2.04 (s) 2.10 (s) 2.13 (s) 2.39 (s)	2.03 (s) 2.03 (s) 2.11 (s) 2.15 (s)

a Measured in CDCl$_3$. Chemical shifts (δ) are expressed in parts per million from Me$_4$Si and coupling constants (J) in hertz. b Multiplicity: s = singlet, d = doublet, t = triplet, q = quartet, m = multiplet, br = broad. c Data for C(5)-cinnamoyl ester protons taken from reference (309).

Table 15. ^1H NMR Spectra of Selected Taxoids[a,b]

Protons on	1β-Hydroxybaccatin I (3.56) (181)	1β-Hydroxy-5α-deacetylbaccatin I (3.57) (310)	1β-Acetoxy-5α-deacetylbaccatin I (3.58) (158)	1β-Hydroxy-7β-deacetoxy-7α-hydroxybaccatin I (3.59) (236)
C-1				
C-2	5.45 (d, 4)	5.48 (d, 3.6)	5.50 (d, 4.4)	5.50 (d, 3)
C-3	3.15 (d, 4)	3.18 (d, 3.6)	3.18 (d, 4.4)	3.39 (d, 3)
C-5	4.18 (br t, 3)	4.22 (t, 6)	4.16 (dd, 4.4, 3.0)	5.62 (dd, 12, 4)
C-6α	1.8 (m)	1.76 (ddd, 15, 4, 3)	1.76 (ddd, 15, 4.1, 3.0)	1.78, 1.86
C-6β		2.08 (m)	2.12 (ddd, 15, 9.9, 4.4)	
C-7	5.45 (dd, 11, 5)	4.49 (dd, 9.9, 4.0)	5.52 (dd, 9.9, 4.1)	3.00 (m)
C-9	5.99 (d, 11)	6.05 (d, 11)	6.10 (d, 11.0)	5.99 (d, 11)
C-10	6.19 (d, 11)	6.22 (d, 11)	6.22 (d, 11.0)	6.23 (d, 11)
C-13	6.0 (m)	6.09 (dd, 9.9, 7.2)	6.09 (q, 9.9, 7.2)	5.99
C-14α	2.2 (m)	1.89 (dd, 15, 7.2)	1.96 (dd, 15, 7.2)	1.92 (q, 15)
C-14β	2.5 (dd, 15, 9)	2.54 (dd, 15, 9.7)	2.52 (dd, 15, 9.7)	2.53 (q, 15)
C-16-Me	1.63 (s)	1.64 (s)	1.62 (s)	1.60 (s)
C-17-Me	1.24 (s)	1.23 (s)	1.22 (s)	1.13 (s)
C-18-Me	2.21 (s)[c]	2.26 (s)	2.24 (s)	1.99 (s)
C-19-Me	1.24 (s)	1.24 (s)	1.24 (s)	1.21 (s)
C-20	2.29 (d, 5); 3.53 (d, 5)	2.52 (d, 5); 3.56 (d, 5)	2.52 (d, 5); 3.63 (d, 5)	2.25 (d, 5); 3.63 (d, 5)
OAc	1.98 (s), 2.07 (s); 2.02 (s), 2.10 (s); 2.04 (s), 2.21 (s)	1.98 (s), 2.09 (s); 2.04 (s), 2.12 (s); 2.05 (s)	1.89 (s), 2.18 (s); 2.00 (s), 2.22 (s); 2.18 (s), 2.24 (s)	2.06 (s), 2.11 (s); 2.06 (s), 2.20 (s); 2.09 (s)

[a] Measured in $CDCl_3$. Chemical shifts (δ) are expressed in parts per million from Me_4Si and coupling constants (J) in hertz. [b] Multiplicity: s = singlet, d = doublet, t = triplet, q = quartet, m = multiplet, br = broad. [c] Uncertain because of the six acetate signals.

Table 16. ¹H NMR Spectra of Selected Taxoids [a, b]

Protons on	1β-Dehydroxybaccatin IV (3.72) (182)	1β-Dehydroxy-4α-deacetylbaccatin IV (3.73) (158)	Baccatin VI (3.74) (235)	1β-Dehydroxybaccatin VI (3.72) (182)
C-1				1.98 (dd, 10.9, 1.9)
C-2	5.87 (dd, 5.9, 1.9)	5.65 (d, 5.5)	5.86 (d, 6)	5.85 (dd, 10.9, 5.9)
C-3	2.86 (d, 5.9)	3.03 (d, 5.5)	3.18 (d, 6)	2.98 (d, 5.9)
C-5	4.97 (d, 8.8)	5.00 (d, 10)	4.97 (d, 10)	4.98 (d, 8.6)
C-6α			1.87	1.85 (ddd, 15.1, 9.7, 8.0)
C-6β			2.48	2.51 (dd, 15.1, 8.0)
C-7	5.50 (dd, 9.9, 7.7)	5.52 (dd, 11.0, 6.5)	5.55 (t, 8)	5.55 (dd, 9.7, 8.0)
C-9	5.89 (d, 11.4)	6.03 (d, 12.4)	6.00 (d, 10)	5.98 (d, 11.2)
C-10	6.13 (d, 11.4)	6.16 (d, 12.4)	6.22 (d, 10)	6.18 (d, 11.2)
C-13	5.88 (br t, 8.3)	6.13 (dd, 9.1, 7.2)	6.19	5.92 (br t, 8.3)
C-14α		1.83 (ddd, 15.2, 9.0, 1.4)	2.17	1.64 (dd, 15.1, 8.3)
C-14β		2.46 (ddd, 15.2, 7.2, 1.4)		2.40 (ddd, 15.1, 10.9, 8.3)
C-16-Me	1.77 (s)	1.67 (s)	1.77 (s)	1.85 (s)
C-17-Me	1.11 (s)	1.20 (s)	1.22 (s)	1.11 (s)
C-18-Me	1.95 (s)	1.96 (s)	2.10 (s)	1.89 (s)
C-19-Me	1.52 (s)	1.56 (s)	1.58 (s)	1.56 (s)
C-20	4.18 (d, 8.1) / 4.49 (d, 8.1)	4.20 (d, 7.4) / 4.53 (d, 7.2)	4.13 (d, 8) / 4.34 (d, 8)	4.10 (d, 8.2) / 4.25 (d, 8.2)
OAc	2.01 (s) 2.25 (s) / 2.15 (s) 2.25 (s) / 2.20 (s) 2.25 (s)	1.95 (s) 2.17 (s) / 2.10 (s) 2.20 (s) / 2.10 (s)	1.99 (s) 2.19 (s) / 2.20 (s) 2.28 (s) / 2.10 (s)	2.02 (s) 2.18 (s) / 2.08 (s) 2.26 (s) / 2.10 (s)
C-2-OBz				7.44 (m) / 7.58 (m) / 8.04 (m)

[a] Measured in CDCl₃. Chemical shifts (δ) are expressed in parts per million from Me_4Si and coupling constants (J) in hertz. [b] Multiplicity: s = singlet, d = doublet, t = triplet, q = quartet, m = multiplet, br = broad.

Table 17. ¹H NMR Spectra of Selected Taxoids[a, b]

Protons on	Baccatin III (3.77) (181)	1-Dehydroxybaccatin III (3.80) (310)	13-Oxo baccatin III (181)	19-Hydroxybaccatin III (3.79) (175)	Baccatin V (3.81) (181)
C-1		1.96			
C-2	5.58 (d, 7)	5.62 (d, 8)	5.72 (d, 7)	6.34 (d, 7)	5.71 (d, 7.5)
C-3	3.84 (d, 7)	4.00 (d, 8)	3.95 (d, 7)	3.88 (d, 7)	4.01 (d, 7.5)
C-5	4.94 (dd, 8, 2)	5.00 (dd, 9, 1)	4.97 (dd, 9, 2)	5.0 (m)	4.93 (dd, 5, 3)
C-6α	2.4 (m)	1.86 (m), 2.15 (m)	2.6 (m)	2.6 (m)	2.1 (m)
C-6β					3.67 (dt)
C-7	4.4 (m)	4.47 (dd, 11, 7)	4.48 (dd, 7, 4)	4.4 (m)	
C-10	6.28 (s)	6.32 (s)	6.48 (s)	6.41 (s)	6.82 (s)
C-13	4.82 (br t, 9)	4.88 (br t, 8)		5.0 (m)	4.8 (m)
C-14α	2.4 (m)	2.56 (m)	2.68 (d)[c]	2.6 (m)	2.3 (m)
C-14β			3.05 (d, 20)[c]		
C-16-Me	1.08 (s)	1.26 (s)	1.28 (s)	1.25 (s)	1.11 (s)
C-17-Me	1.08 (s)	1.10 (s)	1.23 (s)	1.11 (s)	1.05 (s)
C-18-Me	2.01 (d, 1)	2.00 (s)	2.12 (s)	2.07 (d, 1.5)	1.99 (d, 1.5)
C-19-Me	1.63 (s)	1.68 (s)	1.72 (s)	4.72[c]	1.63 (s)
C-20α	4.25 (d, 8)	4.30 (d, 8)	4.35 (d, 8)	4.30[d]	4.36 (d, 9)[e]
C-20β	4.11 (d, 8)	4.18 (d, 8)	4.15 (d, 8)		4.35 (d, 9)[e]
OBz (ortho)	8.05 (dd, 8, 2)	8.05 (dd, 8, 2)	8.10 (dd, 8, 2)	8.13 (dd, 8, 2)	8.12 (dd, 8, 2)
OBz (meta)	7.46 (m)	7.46 (m)	7.58 (m)	7.54 (m)	7.52 (m)
OBz (para)					
OAc's	2.20 (s)	2.22 (s)	2.22 (s)	2.26 (s)	2.20 (s)
	2.24 (s)	2.37 (s)	2.33 (s)	2.29 (s)	2.35 (s)

[a] Measured in $CDCl_3$. Chemical shifts (δ) are expressed in parts per million from Me_4Si and coupling constants (J) in hertz. [b] Multiplicity: s = singlet, d = doublet, t = triplet, q = quartet, m = multiplet, br = broad. [c] AB quartet (J = 12). [d] AB quartet (J = 8). [e] Chemical shift values are interchangeable within columns.

Table 18. ^1H NMR Spectra of Selected Taxol Analogues[a, b]

Protons on	Taxol (3.82) (40)	7-epi Taxol (3.83) (40)	2',7-Diacetyltaxol (134)	10-Deacetyl-10-oxo-7-epi-taxol (3.100) (181)	Cephalomannine (3.88) (181)	10-Deacetyl Cephalomannine (3.89) (181)
C-2	5.67 (d, 7.1)	5.76 (d, 7.5)	5.60 (d, 7)	5.86 (d, 7)	5.62 (d, 7)	5.68 (d, 7.5)
C-3	3.79 (dd, 7.0, 1.0)	3.92 (d, 7.5)	3.87 (d, 7)	4.00 (d, 7)	3.74 (d, 7)	3.88 (d, 7.5)
C-5	4.94 (dd, 9.6, 2.3)	4.91 (dd, 9.0, 3.5)	4.90 (d, 9)	4.90 (dd, 7, 5)	4.87 (dd, 7, 2)	4.92 (dd, br d, 9)
C-6α	2.54 (ddd, 14.8, 9.7 6.7)	2.33 (dd, 16.1, 9.2, 2.1)	2.2 (m)	2.3 (m)	2.0 (m)	2.0 (m)
C-6β	1.88 (ddd, 14.7, 11.0, 2.3)	2.27 (dd, 16.0, 5.0, 3.7)				
C-7	4.40 (ddd, 10.9, 6.7, [4.3]c)	3.70 (br d)	5.53 (m)	3.84 (dt, 10, 3)	4.32 (m)	4.2 (m)
C-10	6.27 (s)	6.80 (s)	6.18 (s)		6.23 (s)	5.18 (s)
C-13	6.23 (tq, 9.0, 1.5)	6.23 (tq, 9.0, 1.5)	6.15 (t, 8)	6.19 (br t, 8)	6.15 (br t, 9)	6.17 (br t, 7.5)
C-14α	2.35 (dd, 15.4, 9.0)	2.42 (d, 15.5, 9.3)	2.55 (m)			
C-14β	2.28 (ddd, 15.3, 9.0, 0.6)	2.25 (d, 15.4, 9.1)		2.3 (m)	2.5 (m)	2.4 (m)
C-16-Me	1.14 (s)	1.15 (s)	1.14 (s)	1.17 (s)	1.24 (s)	1.22 (s)
C-17-Me	1.24 (s)	1.19 (s)	1.09 (s)	1.09 (s)	1.12 (s)	1.12 (s)
C-18-Me	1.79 (d, 1.5)	1.79 (d, 1.5)	1.74 (br s)	1.78 (s)	1.78 (br s)	1.78 (d, 1.5)
C-19-Me	1.68 (s)	1.67 (s)	1.91 (s)	1.70 (s)	1.65 (s)	1.68 (s)
C-20α	4.30 (ddd, 8.4, 1.1, 0.8)	4.39 [AB, 8.7, J_{CH} = 155]	4.25 (d, 8)	4.34 (d, 8)	4.24 (d, 8)	4.32 (d, 9)
C-20β	4.19 (dd, 8.5, 1.0)		4.11 (d, 8)	4.40 (d, 8)	4.12 (d, 8)	4.18 (d, 9)
C-2'	4.78 (dd, [5.4]c, 2.7)	4.81 (d, 2.6)	5.50 (d, 3)	4.80 (d, 3)	4.66 (d, 3)	4.69 (d, 3)
C-3'	5.78 (dd, 8.9, 2.8)	5.81 (dd, 9.0, 2.5)	5.89 (dd, 9, 3)	5.77 (dd, 9, 3)	5.56 (dd, 9, 3)	5.60 (dd, 9, 3)
3'-NH	7.01 (d, 8.9)	7.00 (d, 9.1)	6.83 (d, 9)	6.95 (d, 9)	6.52 (d, 9)	6.62 (d, 9)

Table 18. (Contd)

Protons on	Taxol (3.82) (40)	7-epi Taxol (3.83) (40)	2',7-Diacetyltaxol (134)	10-Deacetyl-10-oxo-7-epi-taxol (3.100) (181)	Cephalomannine (3.88) (181)	10-Deacetyl Cephalomannine (3.89) (181)
OBz (ortho)	8.13 (dd, 8.4, 1.3)	8.18 (dd, 8.5, 1.3)	8.06 (dd, 7, 1)	8.15 (d, 7)	8.05 (dd, 8, 2)	8.12 (dd, 8, 2)
OBz (meta)	7.51 (cm)	7.34–7.56 (m)	7.45 (t, 7)		} 7.46 (m)	} 7.56 (m)
OBz (para)	7.61 (tt, 7.4, 1.4)	7.62 (tt, 7.4, 1.6)	7.52 (t, 7)			
3' Ph (ortho)	7.48 (cm)	} 7.34–7.56, (m)	} 7.3 (m)	} 7.4 (m)	} 7.34 (s)	} 7.38 (s)
3' Ph (meta)	7.42 (cm)					
3' Ph (para)	7.35 (tt, 7.3, 1.6)					
NBz (ortho)	7.74 (dd, 3.3, 1.2)	7.72 (dd, 8.4, 1.3)	7.67 (dd, 7, 1)	7.68 (d, 7)		
NBz (meta)	7.40 (cm)	} 7.34–7.56 (m)	7.45 (t, 7)d	} 7.4 (m)		
NBz (para)	7.49 (cm)		7.67 (dd, 7, 1)d			
1-OH	1.98 (bs)					
7-OH	2.48 (br s [d(4.4)]c)	4.68 (br s)		4.43 (d, 10)		
2'-OH	3.61 (br s [d(5.4)]c)	2–2.5		3.61 (br s)		
4-OAc	2.38 (s)	2.50 (s)	e	2.44 (s)	2.32 (s)	2.35 (s)
10-OAc	2.23 (s)	2.19 (s)	e		2.20 (s)	
C-2''-Me					1.78 (br s)	1.75 (s)
C-3''					6.37 (d, 7)	6.43 (br dd, 7, 3)
C-4''-Me					1.70 (d, 7)	1.7f

a Measured in $CDCl_3$. Chemical shifts (δ) are expressed in parts per million from Me_4Si and coupling constants (J) in hertz. b Multiplicity: s = singlet, d = doublet, t = triplet, q = quartet, m = multiplet, cm = complex multiplet, br = broad. c [ ] denotes additional data from a different sample with slower exchange of hydroxyls. d Chemical shift values are interchangeable within columns. e Four acetate signals at 1.96 (s), 2.08 (s, 6H), 2.36 (s). f Not resolved.

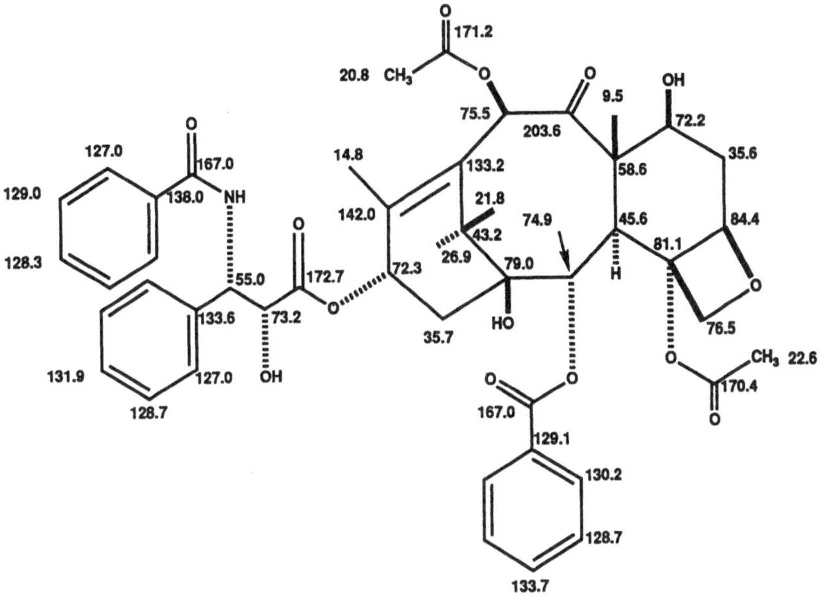

Fig. 4. ^{13}C-NMR assignments for taxol.

4.2.5. Mass Spectrometry

Little information on the mass spectrometry of taxoids has been reported other than molecular ions of new compounds. Although it was stated several years ago (*151*) that "the low volatility of taxinine derivatives made them unsuitable for mass spectroscopic measurements", modern instrumentation can in principle overcome this problem, especially using FAB or some other soft ionization technique to produce molecular ions and then collisional activation to produce daughter ion spectra (MS/MS). Indeed, such an approach to the mass spectrum of taxol has been described in one abstract (*176*), but a full account of this work has yet to appear. Similar studies are reported by Cooks and his collaborators (*104*).

Some information on the fragmentation of the taxoids is scattered in the literature. Thus the mass spectrum of the dihydrotaxinine derivative tetraacetyltaxinol (**4.11**) shows peaks due to sequential loss of acetic acid, but it also shows intense peaks at m/z 137, 135, and 107 (*151*). The peak at m/z 137 is characterized as a ring A fragment, while those at m/z 135 and 107 arise from ring C. Taxoids with the more usual Δ^{11}-double bond do not show corresponding peaks, however; thus the spectra recorded for a taxoid such as (**4.12**) failed to show significant ions at m/z 135 and 107

Table 19. ^{13}C NMR Spectra of Selected Taxoids [a]

Carbon	Taxusin[b] (3.1) (52)	Taxine B[b] (3.29) (70)	Brevifoliol[c] (3.16) (2)	Austrospicatine[b] (3.19) (68)	2'β-Deacetyoxy-austrospicatine[b] (3.24) (68)	2α,13α-Dihydroxy 9α,10β-diacetoxy-5α-cinnamoxy-4(20),11-diene[b] (3.31) (311)
C-1	40.5	77.9	76.1	40.6	40.7	51.08
C-2	32.0	71.6	29.5	27.6	27.45	70.30
C-3	33.1	46.8	38.3	38.1	37.95	45.71
C-4	149.0	144.0	151.8	145.9	146.85	144.09
C-5	76.3	78.2	72.8	76.2	75.55	70.58
C-6	27.4	29.1	36.5	34.1	34.35	26.10
C-7	28.4	26.4	70.3	68.9	70.3	28.34
C-8	43.0	45.6	62.9	39.5	46.4	37.37
C-9	77.6	75.3	77.6	76.8	77.15	75.93
C-10	72.6	76.6	70.8	71.4	71.95	78.82
C-11	135.1	153.5	149.8	134.3	135.50	136.99
C-12	137.2	139.1	134.4	137.2	137.55	134.02
C-13	70.8	199.9	77.4	70.4	70.75	75.93
C-14	27.4	44.5	47.6	31.3	31.6	29.49
C-15	39.3	42.3	45.4	46.1	39.6	44.97
C-16	27.4	20.5	27.2	27.3	27.6	18.18
C-17	31.1	34.2	25.0	30.8	31.25	27.18
C-18	14.8	13.9	20.9	15.6	15.2	31.63
C-19	17.8	17.9	13.0	13.6	13.45	15.29

	1	2	3	4	5	6
C-20	114.1	118.1	112.0	116.9	116.05	119.47
C-1'		171.3			170.75	166.41
C-2'		38.3		74.4	39.5	118.78
C-3'		66.5		70.4	67.2	145.43
C-1''		138.5	129.9	133.9	135.0	130.55
C-2'' and 6''		128.2	129.6	128.2	128.7	128.05
C-3'' and 5''		128.7	129.0	130.2	128.7	129.03
C-4''		127.6	133.6	128.2	128.1	134.27
$N(Me)_2$		42.3		43.2	42.1	
$OCOCH_3$	20.8 21.0 21.4 21.8	21.2	12.2 21.6	21.4 20.9 × 3 20.3	21.4 21.4 21.0 20.85	21.02 21.32
$OCOCH_3$	170.1 170.1 170.4 170.6	170.2	170.4 170.6	170.7 169.8 170.2 169.2 169.9 167.7	170.65 169.25 170.35 169.55	170.40 170.48

[a] Chemical shifts are reported in δ units from internal standard TMS. [b] Solvent: $CDCl_3$. [c] Solvent: CD_2Cl_2.

Table 20. ^{13}C NMR Spectra of Selected Taxoids[a,b]

Carbon	Baccatin I (3.54) (219)	5α-Deacetyl-baccatin I (3.55) (219)	1β-Hydroxy-baccatin I (3.56) (219)	1β-Hydroxy-5α-deacetyl-baccatin I (3.57) (310)	Spicataxine (3.63) (69)	Nicaustrine[e] (3.65) (69)
C-1	47.9	47.4	75.2	76.0	47.8	47.6
C-2	69.0	69.1	72.1	70.8	71.55	71.35
C-3	39.2	37.4	41.4	41.4	38.45	38.75
C-4	58.7	61.0	58.3	58.3	59.5	59.35
C-5	77.7[c]	75.8[c]	77.7[c]	68.8	79.95	78.65
C-6	31.3	33.1	31.1	31.1	24.15	23.95[c]
C-7	75.5[c]	76.3[c]	76.0[c]	71.2	24.8	26.7[c]
C-8	46.5	47.1	46.8	46.7	43.4	43.4
C-9	71.2	71.8	71.1	75.2	76.05	77.8
C-10	70.2[d]	69.8	70.8	77.8	75.5	71.75
C-11	134.4	135.5	135.6	140.4	135.05	134.5
C-12	138.0	139.0	140.2	135.7	137.25	138.25
C-13	70.0[d]	69.8	68.7	72.2	70.25	70.4
C-14	28.9	29.2	38.6	38.8	28.55	28.6
C-15	38.4	38.1	43.3	43.3	38.7	38.35
C-16	27.1	26.2	21.9	15.4	27.1	26.6
C-17	31.3	32.0	28.5	21.9	31.2	31.5

C-18	15.3	16.0	15.4	28.4	15.15	15.45
C-19	13.8	13.3	13.7	13.7	19.5	19.25
C-20	49.7	49.9	49.9	49.9	50.2	50.45
C-1′					171.1	170.8
C-2′					38.95	39.2
C-3′					66.75	66.85
C-1″					not identified	not identified
C-2″ and 6″					128.4	128.6
C-3″ and C-5″					128.3	128.6
C-4″					127.85	128.6
N(Me)$_2$					42.3	42.3
OCOCH$_3$	20.6, 20.9, 21.0 / 21.4, 21.4, 21.6	20.7, 20.9, 21.0 / 21.4, 21.4	20.6, 20.8, 21.1 / 21.4, 21.4, 21.6	20.7, 20.9, 21.4 / 21.4, 21.7, b.	21.9, 21.2, 21.2	22.05, 22.3, 20.9
OCOCH$_3$	168.0, 168.8, 169.0 / 169.5, 169.5, 169.9	168.5, 169.3, 170.0 / 169.5, 169.5, 169.9 / 170.0, 170.0	169.0, 169.0, 169.6 / 169.6, 169.8, 169.8	170.1, 169.8, 169.3 / 169.2, 169.1	170.2, 169.45, 168.75	169.9, 169.5, 168.5

[a] Chemical shifts are reported in δ units from internal standard TMS. [b] Solvent: $CDCl_3$. [c,d] Chemical shifts in the same column are interchangeable. [e] C-9 Nicotinoyl carbons: C-2‴ (153.9), C-3‴ (125.7), C-4‴ (137.2), C-3a‴ (164.65), C-5‴ (123.6), C-6‴ (151.05).

Table 21. ^{13}C NMR Spectra of Selected Taxoids [a, b]

Carbon[c]	Taxol (3.82) (40)	7-epi Taxol (3.83) (40)	Cephalomannine[d] (3.88) (40)	Baccatin III (3.77) (219)	Baccatin IV (3.67) (219)	Baccatin V (3.78) (219)
C-1	79.0	79.1	79.0	75.3	75.2	75.4
C-2	74.9	75.3	74.9	79.2	78.8	79.1
C-3	45.6	40.3	45.5	46.3	47.4	40.7
C-4	81.1	82.1	81.1	81.0	81.4	81.9
C-5	84.4	82.8	84.4	84.7	84.1	82.7
C-6	35.6	36.1	35.5	38.8	34.5[e]	39.0
C-7	72.2	75.7	72.17	72.3	72.9	75.8[e]
C-8	58.6	57.6	58.5	58.8	45.8	57.7
C-9	203.6	207.4	203.8	204.4	71.9	207.9
C-10	75.5	78.1	75.6	76.6[e]	69.6	77.7[e]
C-11	133.2	133.5	133.1	132.1	133.8	132.1
C-12	142.0	139.7	142.4	146.6	141.4	144.4
C-13	72.3	72.3	72.24	68.0	70.4	67.6
C-14	35.7	35.3	35.5	35.7	35.1[e]	35.4
C-15	43.2	42.6	43.1	42.8	42.9	42.2
C-16	21.8	21.2	21.8	20.9	22.3	20.4
C-17	29.6	25.9	26.8	27.0	28.3	26.4
C-18	14.8	14.7	14.7	15.6	15.0	15.6
C-19	9.5	16.1	9.5	9.5	12.7	16.3
C-20	76.5	77.6	76.5	76.4[e]	76.5	75.8[e]
C-1'	172.7	172.9	172.9	—	—	—
C-2'	73.2	73.1	73.3	—	—	—
C-3'	55.0	54.8	54.8	—	—	—

4-OCOCH$_3$	22.6	22.5	22.5	22.6	22.6	f
10-OCOCH$_3$	20.8	20.8	20.8	20.9	20.9	f
4-OCOCH$_3$	170.4	172.4	170.4	170.9	169.8	g
10-OCOCH$_3$	171.2	169.5	171.4	171.6	172.6	g
2-COC$_6$H$_5$	167.00	167.3	167.1	167.3	167.2	
2-COC$_6$H$_5$(q)	129.1	133.7	129.15	129.6	129.7	
2-COC$_6$H$_5$(o)	130.2	130.3	130.3	128.8	128.8	
2-COC$_6$H$_5$(m)	128.71	128.4^e	128.3	130.3	130.2	
2-COC$_6$H$_5$(p)	133.7	133.8	133.8	133.9	133.7	
3'-C$_6$H$_5$(q)	133.6	138.1	131.3	—	—	
3'-C$_6$H$_5$(o)	127.03	127.0	127.0	—	—	
3'-C$_6$H$_5$(m)	128.68	128.8^e	128.8	—	—	
3'-C$_6$H$_5$(p)	131.9	128.9	129.03	—	—	
N-COC$_6$H$_5$	167.02	167.2	—	—	—	
N-COC$_6$H$_5$(q)	138.0	129.4	—	—	—	
N-COC$_6$H$_5$(o)	127.04	127.1	—	—	—	
N-COC$_6$H$_5$(m)	129.0	129.1^e	—	—	—	
N-COC$_6$H$_5$(p)	128.3	132.0	—	—	—	

^{a} Chemical shifts are reported in δ units from internal standard TMS. ^{b} Solvent: CDCl$_3$. ^{c} q = quaternary; o = ortho; m = meta; p = para. ^{d} Tigloyl Carbons: C-1″ (169.1), C-2″ (138.2), C-3″ (132.0), C-4″ (13.9), C-2″ CH$_3$ (12.3). ^{e} Chemical shifts in the same column are interchangeable. ^{f} Acetate signals: 20.8, 20.9, 21.2, 21.4, 21.4, 22.8. ^{g} Acetate signals: 169.1, 169.5, 170.1, 170.4, 170.6, 172.0.

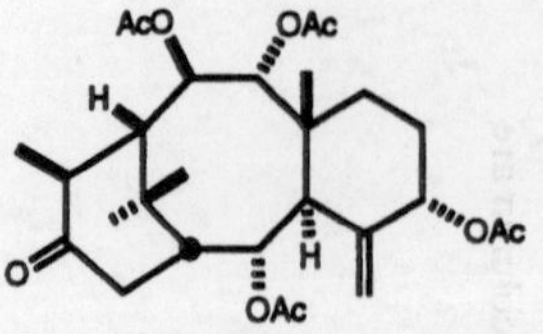

4.11 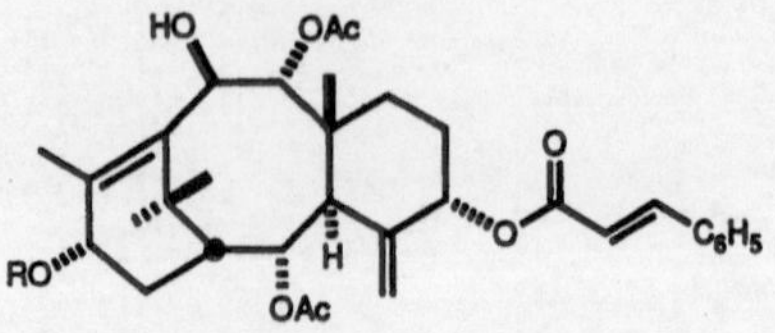   **4.12**

(*311*), even though these ions could be formed by the same mechanism as that proposed for tetraacetyltaxinol (**4.11**).

The original discoverers of taxol reported a molecular ion at m/z 853; although not explicitly stated, this was presumably obtained under electron impact conditions (*285*). Later workers report only fragment ions under EI conditions (*181*), but the FAB mass spectrum of taxol shows intense molecular ions at MH^+, MNa^+, and MK^+. Fragment ions are observed corresponding to the loss of acetic acid and to the loss of the protonated C-13 side-chain (*166*).

4.2.6. X-ray Crystallography

A number of taxoids have been subjected to X-ray crystallographic analysis. Ironically, one of the first reported crystal structures is not that of a natural taxoid but that of the p-bromobenzoate derivative of the rearrangement product (**4.13**) (*32*). As noted earlier, the absolute configuration of taxinine (**1.3**) was established by X-ray analysis of the 14-bromo derivative of taxinol tetraacetate (**2.1**) (*248, 249*).

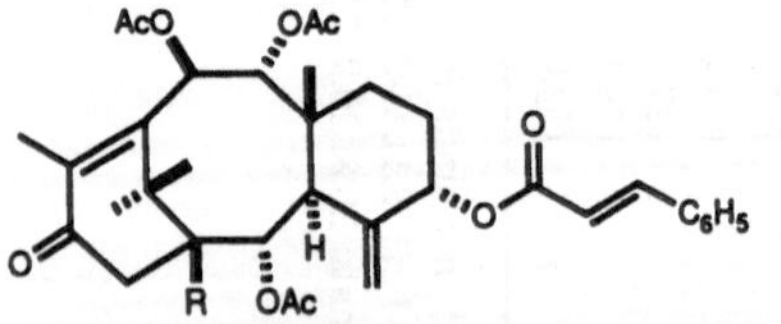

1.2 R = OH
1.3 R = H

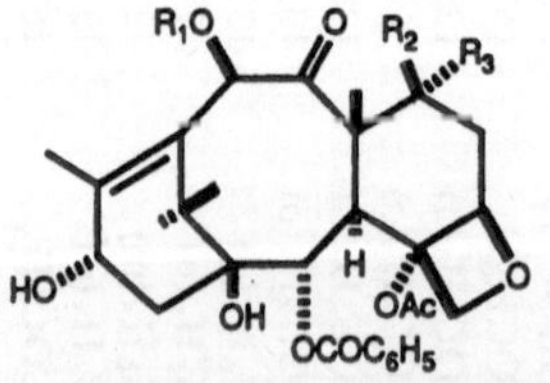

2.1

1.9 R_1 = H, R_2 = OH, R_3 = H
1.10 R_1 = Ac, R_2 = OH, R_3 = H
1.11 R_1 = Ac, R_2 = H, R_3 = OH

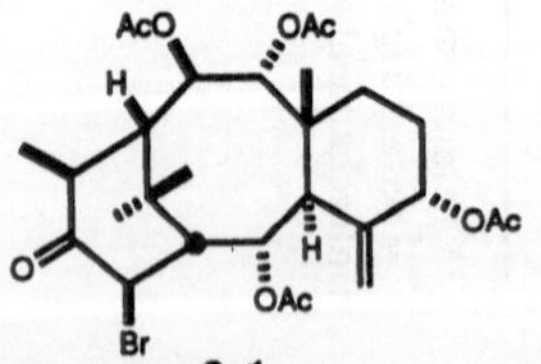

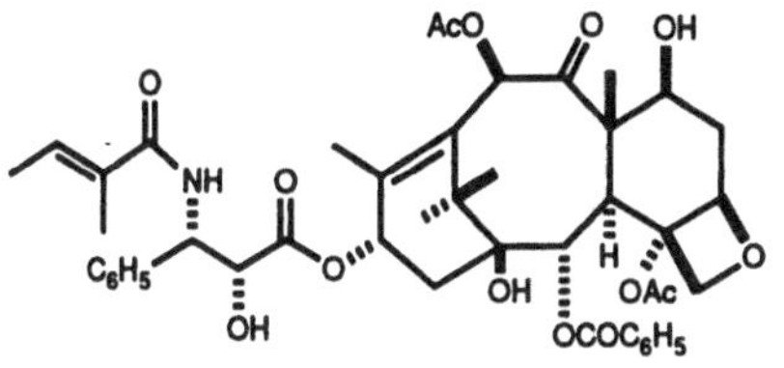

4.1 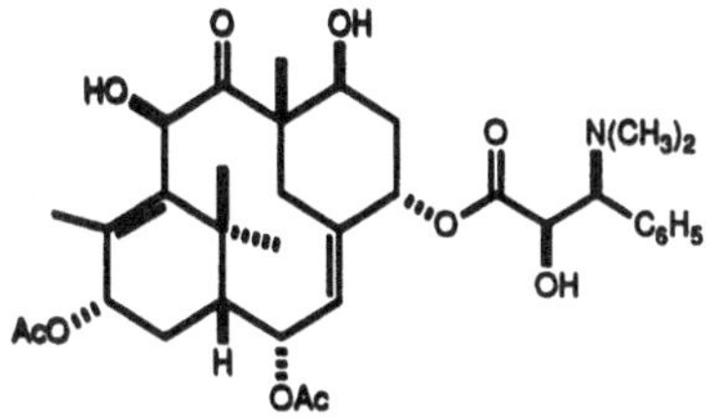   **1.4**

Other taxane diterpenoids which have been analyzed by X-ray crystallography include baccatin V (**1.11**) (*30, 47*), O-cinnamoyltaxicin-l triacetate (**1.2**) (*6*), taxusin (**4.14**) (*102*), taxagifine (**4.15**) (*34*), and taiwanxan (**4.16**) (*103*).

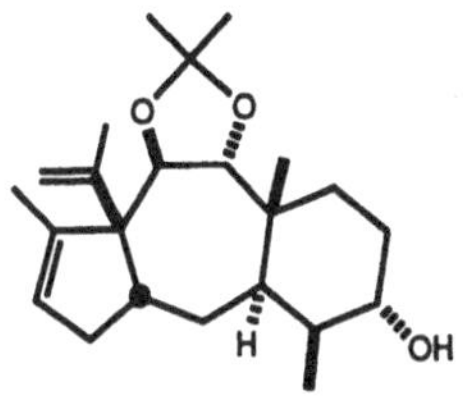

4.13 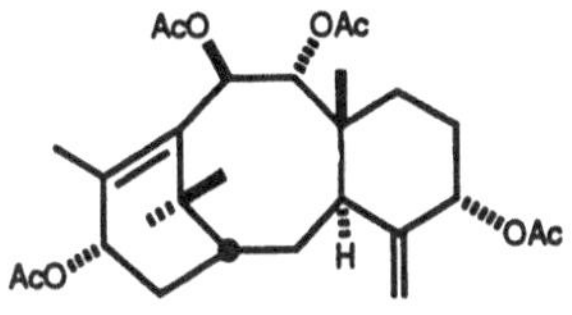   **4.14**

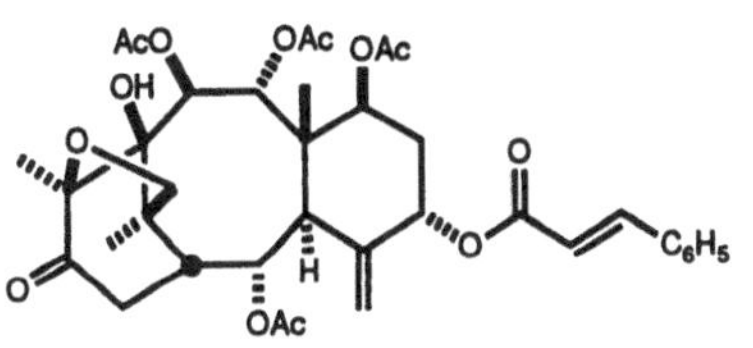

4.15 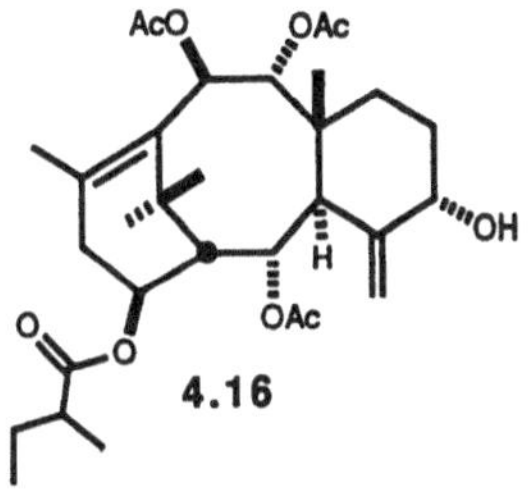   **4.16**

Surprisingly, the X-ray structure of taxol itself has never been determined. The original structure work was carried out on crystalline derivatives of a hydrolysed product. Hydrolysis of taxol gave an N-benzoyl-β-phenylisoserine methyl ester, which was analyzed as its p-bromobenzoate derivative, and the tetraol (**1.9**), which was analysed as its 7,10-bisiodoacetate (*285*). Unfortunately the full details of this work have never been published. However, the X-ray structure of the closely related semisynthetic compound taxotere (**4.17**) has been published (*90*).

The side-chain ester of cephalomannine (**4.1**) was characterized by X-ray crystallography (*181*) and the side-chain ester of taxol has recently been similarly characterized (*210*). Finally, the structure of the unusual

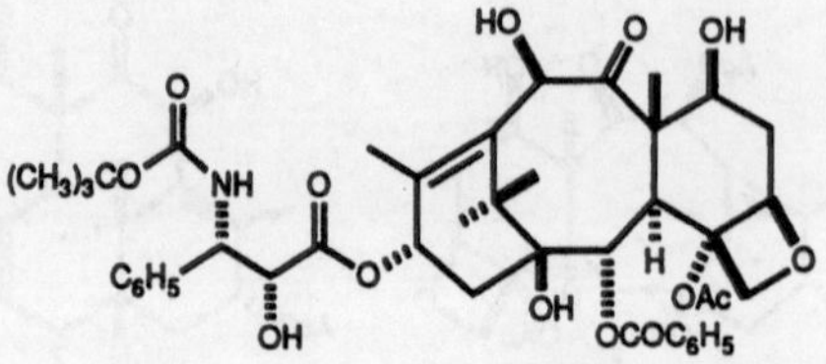

4.17

Taxus alkaloid taxine **A** (**1.4**) has been determined by X-ray crystallography (*83*).

4.3. Chemical Reactivity

4.3.1. *Acylation and Other Protective Group Chemistry*

The taxoids typically have several hydroxyl groups, either free or as the acetate esters, so acylation and deacylation reactions have played an important part in their chemistry.

In the case of taxoids with free hydroxyl groups at the 9- and 10-positions, the 10-hydroxyl group is somewhat more reactive than at the 9-position. Thus the methyl ether (**4.18**) gave 17% of the 9-monoacetate and 40% of the 10-monoacetate on acetylation with acetic anhydride and pyridine (*97*).

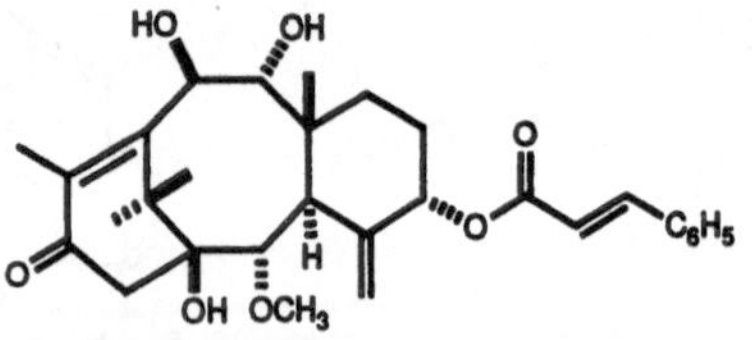

4.18

Acylation of baccatin-III (**1.10**) and 10-deacetylbaccatin-III (**1.9**) is important as a preliminary to the conversion of these compounds to taxol since 10-deacetylbaccatin-III is relatively readily available from *T. baccata* leaves (*35, 55*). Although an early study suggested that acetylation of baccatin III gave the 13-acetyl derivative (*181*), two groups have shown that the 13-hydroxyl group is in fact much less reactive than that at C-7 (*89, 168, 236*). This fact makes the attachment of an appropriate side-chain at C-13 to convert baccatin III to taxol a challenging task; the solutions that have been developed to this problem will be discussed in Sect. 5.3.

1.9 $R_1 = H$, $R_2 = OH$, $R_3 = H$
1.10 $R_1 = Ac$, $R_2 = OH$, $R_3 = H$
1.11 $R_1 = Ac$, $R_2 = H$, $R_3 = OH$

Since 10-deacetylbaccatin III (**4.3**) is probably more readily available than baccatin III, its acetylation becomes of major importance. A study carried out by POTIER and his collaborators (*89*) showed that under mild conditions (24 hr, 20 °C) the 7-acetate (**4.19**) and the 7,10-diacetate (**4.20**) were formed in equal amounts. Under more forcing conditions (48 hr, 60 °C) the 7,10-diacetate (**4.20**) and the 7,10,13-triacetate (**4.21**) were formed in equal amounts, while under the most vigorous conditions (24 hr, 80 °C) only the triacetate was formed. The order of reactivity for acetylation is thus 7 > 10 ≫ 13. The use of more bulky protective groups permitted discrimination between the C-7 and C-10 hydroxyl groups; thus triethylsilylation of 10-deacetylbaccatin-III under carefully optimized conditions gave 7-triethylsilyl-10-deacetylbaccatin-III (**4.22**) in about 85% yield and this could then readily be converted to 7-triethylsilylbaccatin-III (**4.23**) (*55*). The t-butyldimethylsilyl group could not be cleanly introduced onto the C-7 position (*55*).

4.3 $R_1 = R_2 = R_3 = H$
4.19 $R_1 = R_2 = H$; $R_3 = Ac$
4.20 $R_1 = H$; $R_2 = R_3 = Ac$
4.21 $R_1 = R_2 = R_3 = Ac$
4.22 $R_1 = R_2 = H$; $R_3 = SiEt_3$
4.23 $R_1 = H$; $R_2 = Ac$; $R_3 = SiEt_3$

Protection of taxoids in the form of cyclic acetal and orthoester derivatives is also possible. Although the C-9 and C-10 hydroxyl groups are formally *trans* related, they readily form an acetonide under appropriate conditions. Thus treatment of 5-cinnamoyltaxicin-I (**4.24**) with acetone and cupric sulfate converts it to the acetonide (**4.25**) (Scheme 4.2). Formation of an orthoester derivative requires firmer conditions: conversion of the taxicin derivative (**1.2**) to the orthoester mixture (**4.26**) required treatment with methyl iodide and silver oxide (*72*).

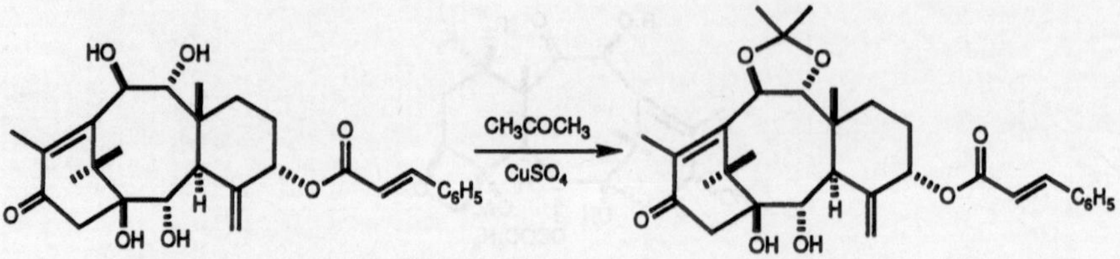

4.24   **4.25**

Scheme 4.2. Conversion of 5-cinnamoyltaxicin-I to its acetonide

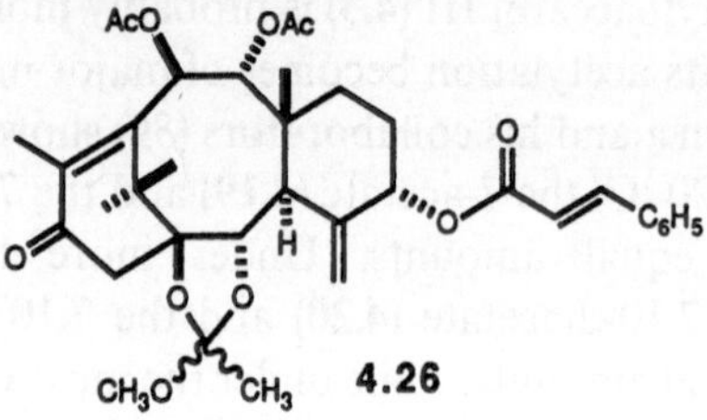

4.26

Acetylation of taxol occurs most readily at the 2'-position, but the 2'-acetate can be further acetylated to the 2',7-diacetate (*178*). This ready acylation of the 2'-position has made it the preferred position for the preparation of prodrugs of taxol; a further advantage of this position is that 2'-esters are more readily hydrolysed than their 7-counterparts. The need for prodrug analogs of taxol stems from its very low water solubility, which makes it difficult to administer by intravenous infusion. Although a surfactant formulation has been developed (*188*), this is not the most desirable mode of administration. A water-soluble prodrug form of taxol that would undergo hydrolysis to taxol in the plasma is thus a highly desirable derivative.

The first prodrugs of taxol to be prepared were 2'-succinyl and 2'-β-alanyl derivatives (*169*). An extensive series of 2'-and 7-succinyl and glutaryltaxol derivatives has also been prepared (*57*), as has a series of 2'- and 7-dimethylaminoacyl derivatives (*174, 252*). The latest in the series of prodrug derivatives is a group of 2'-sulfonate derivatives (*136, 315*). The most active prodrug to date appears to be a 2'-glutarylamino amide derivative of taxol (**4.27**), which is as active as taxol in the B16 melanoma assay; the higher T/C values reported for this compound are an artefact of the way the test was run. It is worth noting that MATHEW and his collaborators used the *t*-butyloxycarbonyl (*t*-BOC) protecting group to prepare their amino acid derivatives and found that clean deprotection of this group could only be achieved by the use of 99% formic acid (*174*).

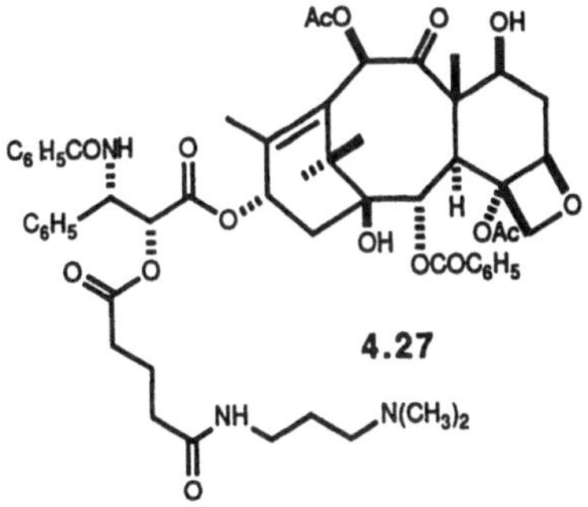

4.27

Protection of the 2'-position of taxol can be achieved with various protecting groups. Although 2'-acetyltaxol can be prepared readily and selectively, its removal is not completely selective as is described below. The use of the chloroacetyl protecting group is convenient, since it is readily removable from the 2'-position by treatment with thioethanolamine or simply with moist silica gel (229). The 2,2,2-trichloroethyloxycarbonyl (troc) derivative has also been used; it is selectively removable by treatment with zinc and acetic acid (37). The troc group cannot, however, be used in the presence of certain non-nucleophilic bases, since it can act as an electrophilic center for intramolecular displacement. Thus treatment of 2',7-di(troc) taxol with 1,5-diazabicyclo[5.4.0]undec-7-ene (DBU) followed by deprotection at C-7 yielded the oxazolone (**4.28**) (169); the same product was formed directly from taxol by treatment with carbonyldiimidazole (57).

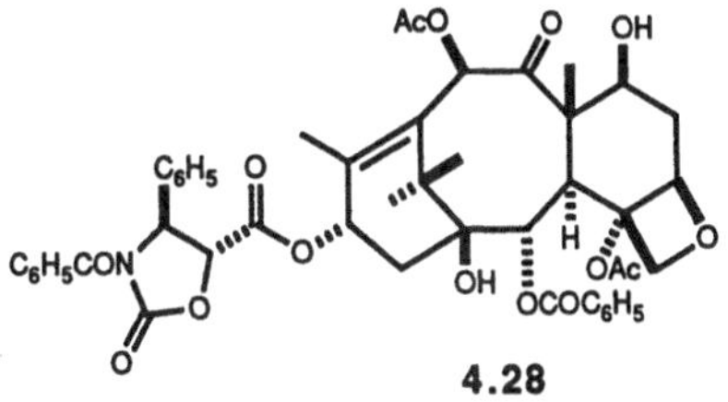

4.28

Ether derivatives of taxol can also be used to advantage. Both 2'-(triethylsilyl)-and 2'-(t-butyldimethylsilyl) taxols have been prepared (166, 169); deprotection can be achieved either with fluoride ion or with dilute acid. A 2'-(1-ethoxyethyl) taxol derivative was also prepared during the synthesis of taxol from baccatin III (55); this protecting group is selectively removed using dilute acid.

4.3.2. Hydrolysis

As noted earlier, taxoids typically carry several acyl substituents, and thus several selective deacylation reactions have been carried out. For simple taxoids such as taxinine (1.3), carrying substituents at C-2, C-5, C-9, and C-10, the order of reactivity to hydrolysis appears to be C-9, C-10 > C-5, C-2. Thus taxinine (1.3) could be selectively hydrolysed to the diol (4.29) (4); this higher reactivity of esters at C-9 and C-10 towards hydrolysis is probably due to simple steric factors, but some element of neighboring group participation could also be a factor. Conversion of the diol (4.29) to the diacetate (4.30) was accomplished by protection of (4.29) as its acetonide derivative, hydrolysis to the 2,5-diol, acetylation, and deprotection (143).

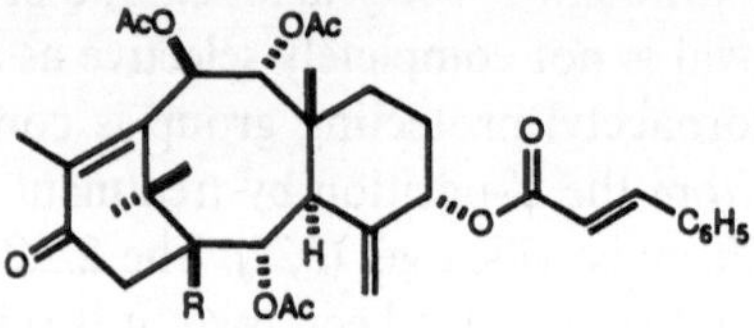

1.2 R = OH
1.3 R = H

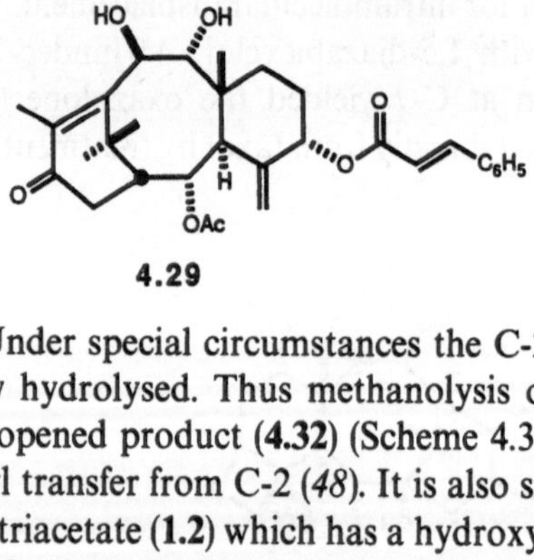

4.29

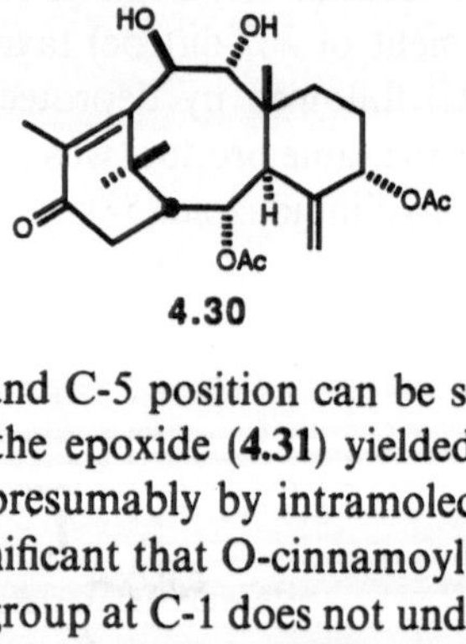

4.30

Under special circumstances the C-2 and C-5 position can be selectively hydrolysed. Thus methanolysis of the epoxide (4.31) yielded the ring-opened product (4.32) (Scheme 4.3), presumably by intramolecular acetyl transfer from C-2 (48). It is also significant that O-cinnamoyltaxicin-l triacetate (1.2) which has a hydroxyl group at C-1 does not undergo the selective hydrolysis observed for taxinine; treatment of (1.2) with cold

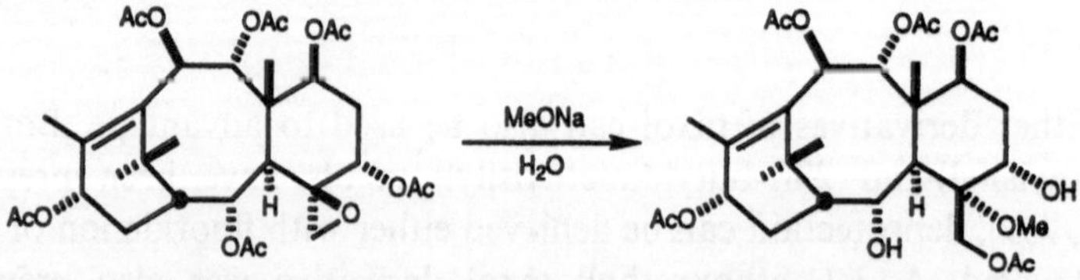

4.31   **4.32**

Scheme 4.3. Hydrolysis of epoxide

methanolic sodium methoxide converts it to O-cinnamoyltaxicin-l (**4.33**) (*4*). Taxicin-l itself has not been isolated, since it decomposes under the alkaline conditions required to produce it (*164*).

Selective hydrolysis of various taxoids isolated from *Austrotaxus spicata* has also been carried out (*68*); in this series the reactivity order is C-13 > C-5 > C7, C-9, C-10. Thus methanolysis of the acetate derivative (**4.34**) yields a mixture of the 13-deacetyl derivative (**4.35**) and the 5,13-deacyl derivative (**4.36**). The greater reactivity of the C-13 acetate in this example is probably related to a neighboring group effect of the C-5 ester amino group.

4.33

4.34

4.35 R$_1$ = COCH(OH)CH(NMe$_2$)C$_6$H$_5$
4.36 R$_1$ = H

Methanolysis of taxol was used as a key reaction in its structure elucidation and yielded the tetraol (**1.9**) together with the methyl ester of the side-chain acid. The structure of the tetraol (**1.9**) was elucidated by X-ray crystallography of its 7,10-bisiodoacetate (*285*). Mild methanolysis of cephalomannine gave a mixture of five taxoid products, identified as baccatin III (**1.10**) (19%), 10-deacetylbaccatin-III (**1.9**) (19%), 10-deacetylcephalomannine (5%), baccatin V (**1.11**) (17%), and 10-deacetyl-baccatin-V (14%); the side chain methyl ester was also obtained in 54% yield (*181*). A study of the hydrolysis of taxol at pH 9 in methanol gave similar results, except that baccatin III was the major hydrolysis product and 10-deacetyltaxol, 7-*epi*taxol, 10-deacetyl-7-*epi*taxol, baccatin-V, 10-deacetylbaccatin-V, and 10-deacetylbaccatin-III were formed in lesser amounts.

The side chain of taxol or cephalomannine can also be removed by a reductive process; the preferred conditions use tetrabutylammonium borohydride in dichloromethane (*168*). Under these conditions the borohydride complexes with the free 2'-hydroxyl group and thus reduces the

side-chain selectively and in high yield; as expected, the reaction fails
when applied to a 2'-acetyl or 2'-silyl derivative. This reaction has proved
useful in a number of situations, and could in principle be used as a key
step in the conversion of cephalomannine to taxol by removal of the side
chain and reacylation of the resulting baccatin III as described in
Sect. 5.3.2.

Selective deacylation of baccatin III has also been investigated in our
group (229). Hydrogenation of baccatin III over Pt yielded a hexahydro
derivative by reduction of the aromatic ring of the C-2 benzoate; the Δ^{11}
double bond was completely unaffected under these forcing conditions.
Mild methanolysis of the 7-triethylsilyl ether of this reduced baccatin III
derivative (4.37) yielded the 10-deacetyl derivative (4.38) and the 4,
10-dideacetyl derivative (4.39) in about equal amounts. Further hydrolysis of the mixture of partially deacylated derivatives converted it to
2-debenzoyl-4,   10-dideacetyl-7-triethylsilylbaccatin-III   (4.40).   The
formation of the 4, 10-dideacetyl derivative was unexpected, and was
shown to be due to neighbouring group participation by the free C-13
hydroxyl group which is relatively close to the C-4 acetate (compare
Fig. 1). Thus methanolysis of 7,13-di(triethylsilyl) hexahydrobaccatin-III
gave *only* the 10-deacetyl derivative under the same conditions as used
previously. Treatment with base for an extended period gave some
additional products in low yield, characterized as the 2,10-dideacyl and
2,4,10-trideacyl-13-de(triethylsilyl) derivatives.

4.38 $R_1 = COC_6H_{11}$; $R_2 = Ac$
4.39 $R_1 = COC_6H_{11}$; $R_2 = H$
4.40 $R_1 = R_2 = H$

4.37

4.3.3. Epimerization at C-7

The C-7 hydroxyl group of taxol and related compounds is part of a
β-hydroxycarbonyl system and is thus subject to epimerization *via* a
retro-aldol/aldol mechanism; this epimerization accompanies essentially
all reactions of taxol and related taxoids under basic conditions. Thus the
mild methanolysis of cephalomannine or taxol described in the previous

section yielded products belonging to both the normal series and the 7-*epi*-series. Epimerization appears to occur more readily in the 10-deacetyl series on standing at room temperature in aqueous methanol. This is probably due to activation of the C-9 carbonyl group by hydrogen bonding with the C-10 hydroxyl group. The driving force for epimerization may be due in part to the formation of a hydrogen bond between the 7-*epi*-hydroxyl group and the C-4 acetate carbonyl group, since this hydrogen bond is seen in the X-ray structure of baccatin V (**1.11**) (*30*). However, the reverse epimerization is also observed (*175*), and thus the stabilization by hydrogen bonding must be balanced by other less favorable interactions.

Epimerization of taxol also occurs under acidic conditions. Thus treatment of taxol with $ZnBr_2$ in methanol at room temperature converts it to a mixture of 10-deacetyltaxol and 10-deacetyl-7-*epi*-taxol, with the 7-*epi*isomer as the major product (*123, 230*). Epimerization has also been reported under neutral conditions in the presence of azobis-(isobutyronitrile) (AIBN) as a free radical initiator (*113*). However, later studies with different batches of AIBN have failed to duplicate this work and it is now suspected that the observed epimerization was due to a trace impurity in the original batch of AIBN used.

4.3.4. Oxidation

For the most part oxidation reactions of the taxoids proceed by normal pathways to yield predictable products, but there are some unusual reactions nonetheless.

Oxidation of double bonds has been restricted to the $\Delta^{4(20)}$ exocyclic methylene group and to the double bond in the cephalomannine side chain; the Δ^{11} double bond is very hindered and is not readily oxidized. Thus reaction of the dihydrotaxicin derivative (**4.41**) with osmium tetroxide converts it to the diol (**4.42**) (*5*), and the conversion of cephalomannine (**4.1**) to its diol derivative (**4.2**) under similar conditions has already been noted (*133*). The Δ^{11} double bond is unaffected by these

4.41   **4.42**

conditions, and also by ozonolysis conditions (*49, 123*). It is, however, oxidized to a small extent by peracids. Thus the taxoid (**4.43**) gives the epoxide (**4.44**) as the major product on treatment with monoperphthalic acid, but smaller amounts of the C-4 epimeric epoxide and of a diepoxide are formed (*48*). Although the diepoxide was apparently not fully characterized, it is most probably a C-4(20), C-11,12 diepoxide.

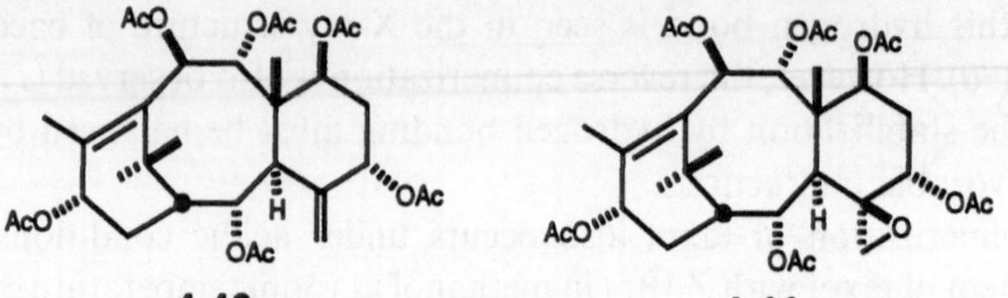

4.43   **4.44**

Oxidation of the C-13 hydroxyl group of taxoids can be accomplished selectively since it is an allylic group. Thus oxidation of the methoxy derivative (**4.45**) with manganese dioxide yields the 13-oxo derivative (**4.46**) (Scheme 4.4); this process is reversible, and reduction of (**4.46**) with sodium borohydride yields (**4.45**) (*97*). In the presence of mildly basic manganese dioxide hydrolysis of a C-13 side chain and oxidation occur in one pot, as exemplified by the conversion of both taxol and baccatin III to 13-oxobaccatin III (**4.47**) on treatment with manganese dioxide (*49, 213, 285*). Under these mildly basic conditions epimerization at C-7 also occurs to give 13-oxobaccatin V (**4.48**) as well

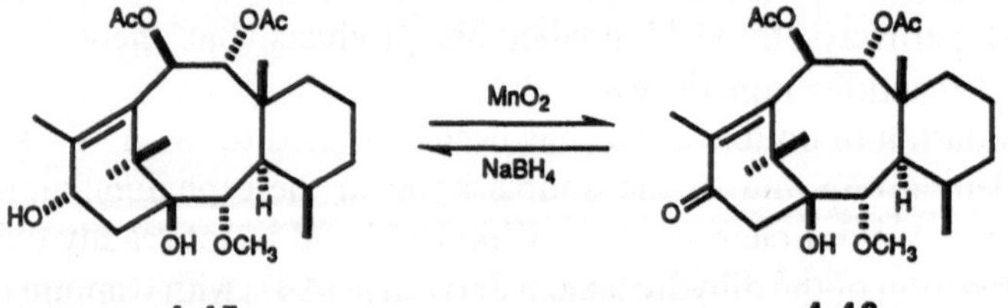

4.45   **4.46**

Scheme 4.4. Oxidation of 13-hydroxy taxoids

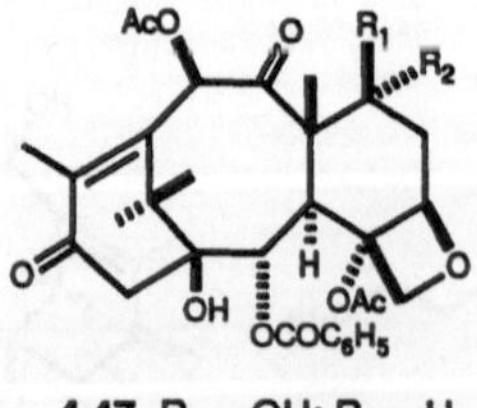

4.47 R$_1$ = OH; R$_2$ = H
4.48 R$_1$ = H; R$_2$ = OH

(181). Oxidation of baccatin III to its 13-oxoderivative also occurs on prolonged treatment with sodium metaperiodate *(50)* and 10-deacetyl-baccatin III is oxidized to its 13-oxo derivative on treatment with chromic acid *(236).*

Oxidation of taxoids with standard oxidizing agents such as chromium trioxide in pyridine or Jones' reagent oxidizes secondary alcohols to ketones in the normal way. Thus oxidation of the taxicin derivatives (**4.49**) and (**4.51**) with chromium trioxide in pyridine gave the corresponding ketones (**4.50**) and (**4.52**) (Scheme 4.5) *(97).* In the case of taxol itself the C-7 hydroxyl group is more readily oxidized than that at C-2′. Oxidation of taxol with Jones' reagent thus yielded initially the C-7 oxo derivative (**4.53**), but prolonged oxidation yielded the 2′,7-dioxo derivative (**4.54**). This latter compound, consistent with the presence of an α-ketoester group, added methanol to form the hemiketal (**4.55**) (Scheme 4.6) *(167).*

Treatment of the 7-oxotaxol (**4.53**) with DBU in dichloromethane at 25 °C or by chromatography on silica gel caused β-elimination to occur leading to the ring D-seco derivative (**4.56**) (Scheme 4.7). Reduction of (**4.56**) (Pt/H$_2$) then gave the unstable ketone (**4.57**) which underwent a retro-Claisen reaction and lactonization to give the lactone (**4.58**) on mild aqueous/alcoholic work-up (Scheme 4.7) *(167).*

More drastic oxidation reactions have been carried out as part of various degradation studies of the taxane ring system. Thus oxidation of taxinine (**1.3**) with selenium dioxide yielded the 14-oxo derivative (**4.59**)

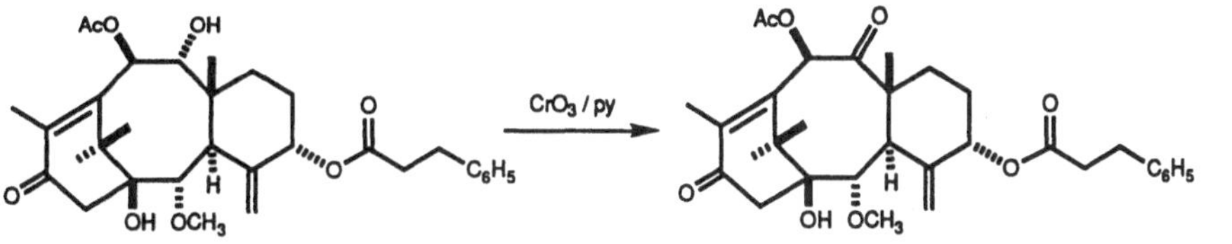

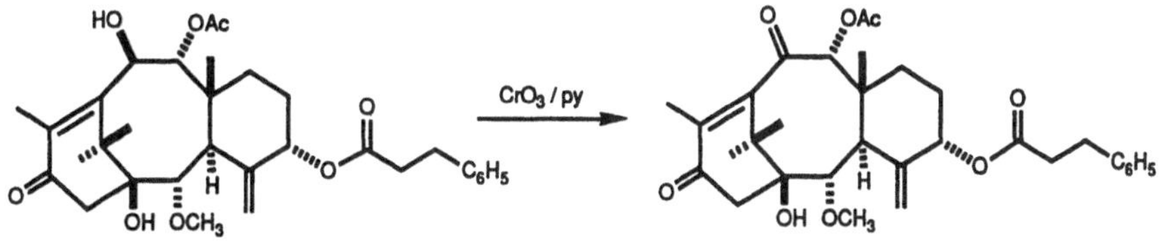

Scheme 4.5. Oxidation of taxicin derivatives

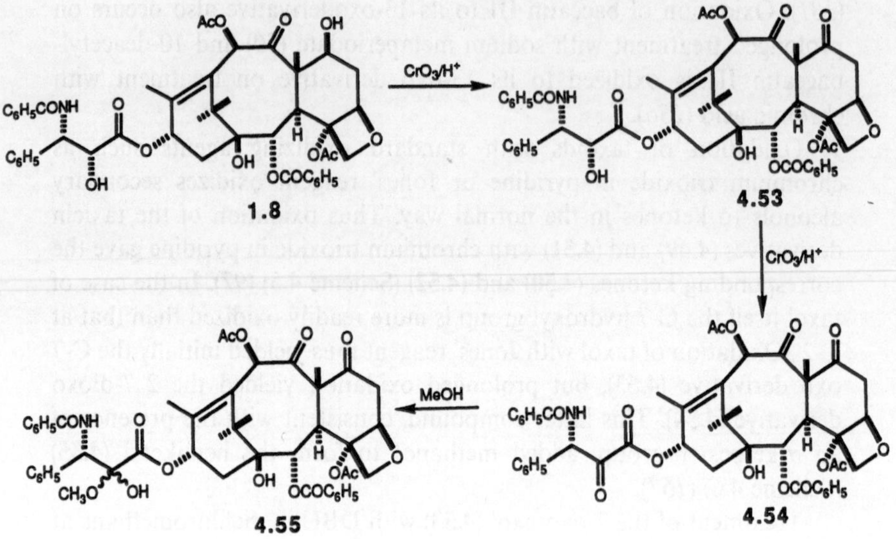

Scheme 4.6. Oxidation of taxol at C-7 and at C-2′

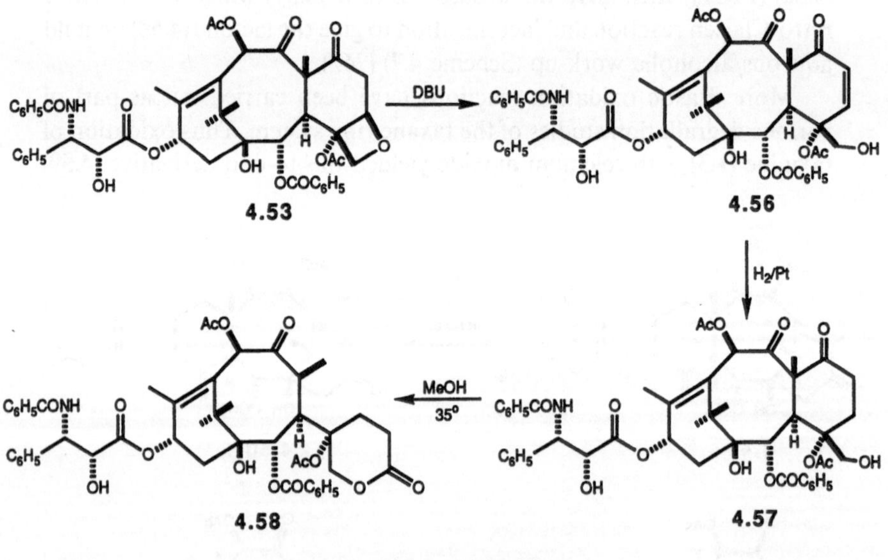

Scheme 4.7. Ring-opening of 7-oxotaxol

and further oxidation of (**4.59**) with hydrogen peroxide followed by acetic anhydride yielded the anhydride (**4.60**) in which elimination of the C-2 acetate had also occurred. Hydrolysis of (**4.60**) then gave the diacid (**4.61**) (Scheme 4.8) (*150, 151*).

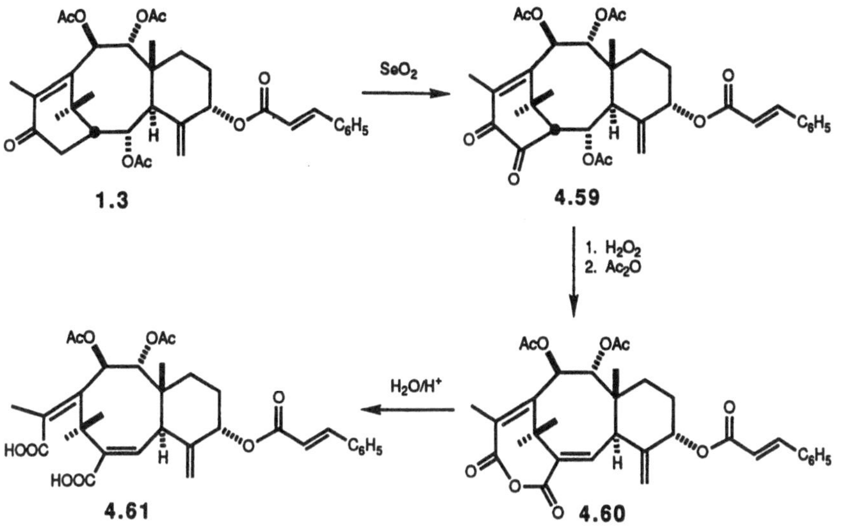

Scheme 4.8. Oxidation of taxinine with SeO$_2$ and H$_2$O$_2$

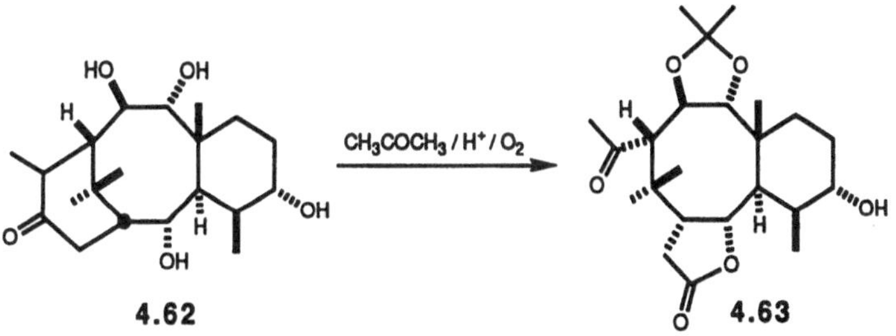

Scheme 4.9. Autoxidation of dihydrotaxinol

An autoxidation product of taxinine was obtained in the course of studies on its chemistry by another Japanese group. Reduction of taxinine with lithium aluminum hydride followed by hydrogenation yielded the dihydrotaxinol (4.62) which could be converted to its isopropylidene derivative on treatment with toluene sulfonic acid in acetone. Prolonged heating under these conditions, however, yielded the product (4.63), presumably formed by autoxidation at the tertiary C-12 position (Scheme 4.9) (170, 280).

4.3.5. Reduction

Simple hydrogenation of taxoids containing a C-4 (20) methylene group proceeds without reduction of the C-11, 12 double bond which is very hindered. Thus reduction of O-β-phenylpropionyltaxicin-1 over palladised charcoal in ethyl acetate yielded mainly the 4, 20-dihydro-compound (4.64) in which the new methyl group is β-oriented (5, 61, 72). When the reaction was carried out in methanol containing 2% water, however, extensive hydrogenolysis of the allylic phenylpropionate group took place and a mixture of (4.64) and the 5-deoxy derivative (4.65) was formed; surprisingly, the deoxy derivative had a stereochemistry at C-4 opposite to that of simple hydrogenation product (4.64).

4.64   4.65

Hydrogenolysis has also been observed in the baccatin III series. Thus hydrogenation of 7-acetylbaccatin-III with Adams catalyst in acetic acid yielded the 13-desoxy derivative (4.66) (50). However, this reaction does not occur on baccatin III itself; instead, hydrogenation of this compound with platinum in acetic acid yielded the hexahydro derivative (4.67) (135, 229).

4.66   4.67

It is noteworthy that hydrogenation of the C-11, 12 double bond did not occur under any of the various hydrogenation conditions described in the previous sections. Reduction of this double bond was observed, however, when the methyl ether diacetate (4.68) of the hydrogenolysis product (4.65) was hydrogenated over platinum in acetic acid. Three hydrogenation products were observed; the alcohol (4.69) formed by reduction of the 13-oxo group and a mixture of diastereomeric dihydro compounds with the probable structures (4.70) and (4.71) (Scheme 4.10) (96). It thus appears that hydrogenation of the C-11,12 double bond can be induced by appropriate functional groups on the taxane skeleton.

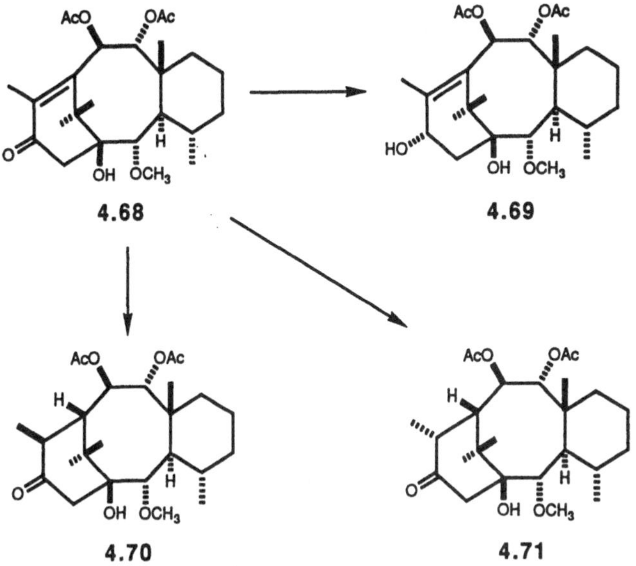

Scheme 4.10. Hydrogenation of taxicin derivatives

Reduction of taxoids with metal hydride reagents gives widely differing products, depending on the metal hydride used. At the simplest level, the 13-oxo group can be reduced with sodium borohydride to the 13-α-alcohol. Thus reduction of the methyl ether (**4.68**) with sodium borohydride gave the alcohol (**4.69**) (*96*), and reduction of 13-oxo-10-deacetyl baccatin III with the same reagent regenerated 10-deacetylbaccatin III (*236*). In the case of taxol, as previously noted, reaction with sodium borohydride (or better, tetrabutylammonium borohydride) leads to a clean reductive cleavage of the side-chain and formation of baccatin III in excellent yield (*168*). Reaction of taxinine (**1.3**) with lithium aluminium hydride in boiling tetrahydrofuran, however, proceeds with an unusual 1,4-reduction to give the dihydro compound taxinol (**4.72**) (*267, 281*).

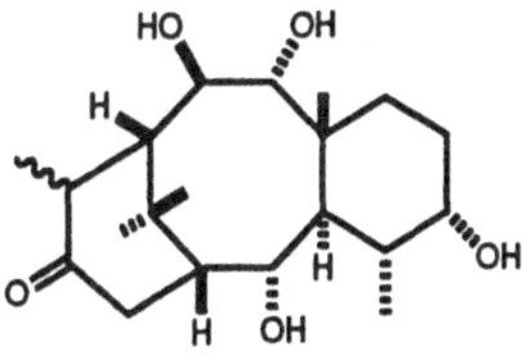

4.72

Reduction of certain taxoids with zinc in acetic acid leads to reductive elimination of the C-10 acetate. Thus reduction of the methyl ether (**4.73**)

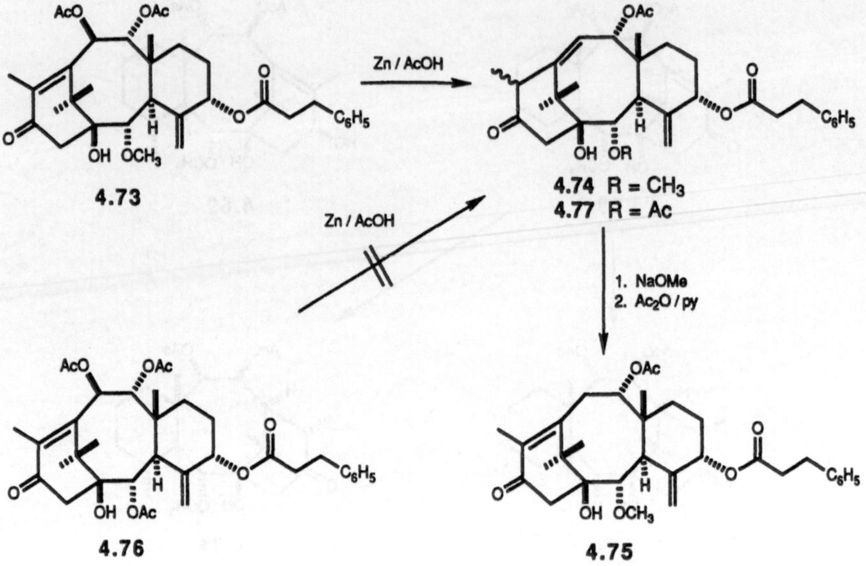

Scheme 4.11. Reduction of taxoids with zinc and acetic acid

with zinc and acetic acid yielded the reduction product (**4.74**); isomeriz-ation of (**4.74**) with sodium methoxide and reacetylation gave the 10-desoxy derivative (**4.75**) (Scheme 4.11) (*96*). A very similar reaction was observed with tetrahydrotaxinine, leading to a 1-desoxy analogue of (**4.74**) (*151*). The reaction took a different course, however, with the dihydrotaxicin derivative (**4.76**); in this case, the expected product (**4.77**) was not formed and instead a different and uncharacterized product was produced with λmax 244nm (Scheme 4.11) (*97*).

4.3.6. *Rearrangements and Related Reactions*

The taxene skeleton is a reasonably stable one, as noted earlier, but it does undergo various rearrangement reactions under certain conditions. Reaction of taxol with triethyloxonium tetrafluoroborate (Meerwein's reagent) under mild conditions followed by aqueous work-up leads to one major product in a mixture with various other substances. This major product was shown to be the 20,O-secotaxol derivative (**4.78**) by a combination of spectroscopic techniques and chemical conversion to its isopropylidene derivative (*230*). Although (**4.78**) is not a skeletal re-arrangement product, its formation must involve neighboring group participation of some nature: one possible mechanism is shown in the

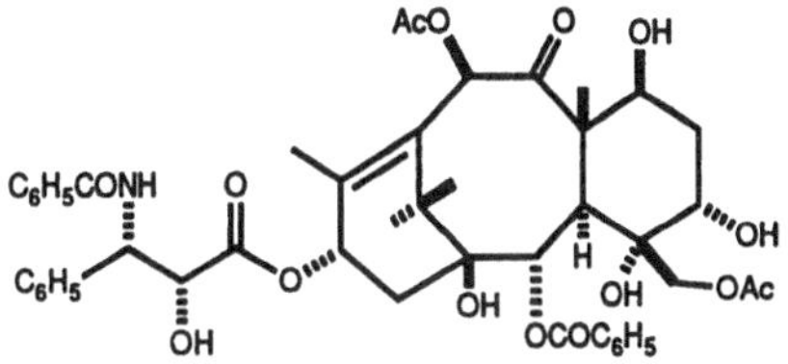

4.78

sequence **4.79** → **4.84** (Scheme 4.12). An orthoester similar to that of one of the intermediates of Scheme 4.12 was proposed in a lecture as the product of the reaction between taxol and zinc bromide (*134*). This structure has since been discarded in favor of a simple conversion of taxol to its 7-epimer under these conditions (*230*).

Conversion of the Meerwein product (**4.78**) to its isopropylidene derivative with acetone in the presence of acid also brought about a skeletal rearrangement of the A-ring and the formation of product (**4.85**). This rearrangement is facilitated by the prior opening of the oxetane ring since taxol itself does not undergo any rearrangement under comparable conditions (*230*). A similar rearrangement is seen, however, when taxol is refluxed with acetyl chloride; in this case the product is the triacetate (**4.86**). The same triacetate is also formed on treatment of (**4.78**) with acetyl chloride, although in this case the conditions required are much milder.

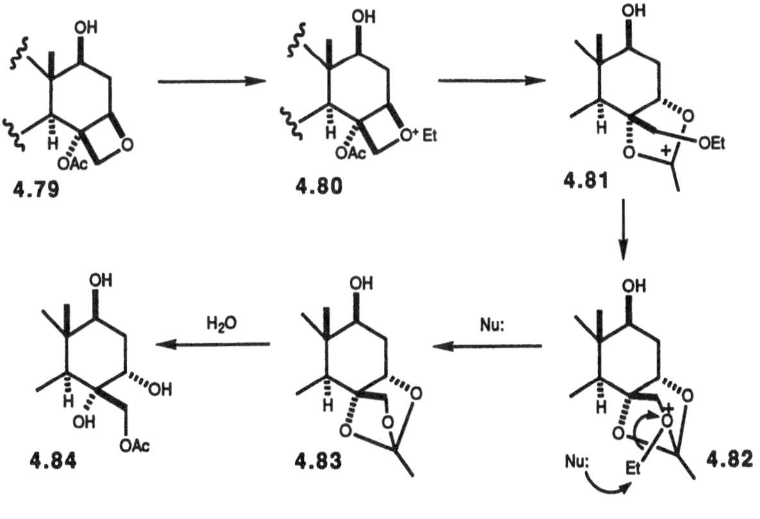

Scheme 4.12. Mechanism of formation of the Meerwein product

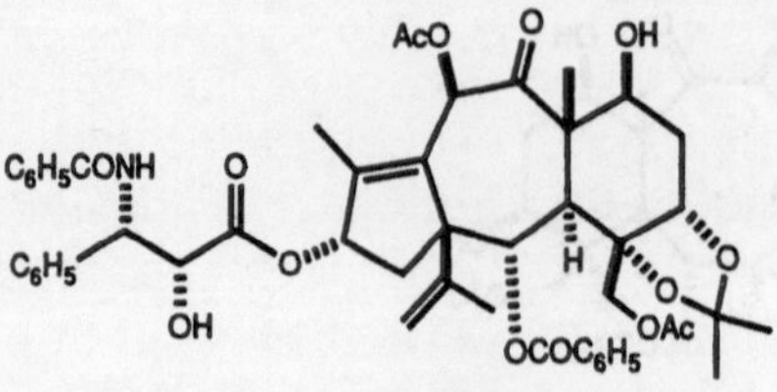

4.85

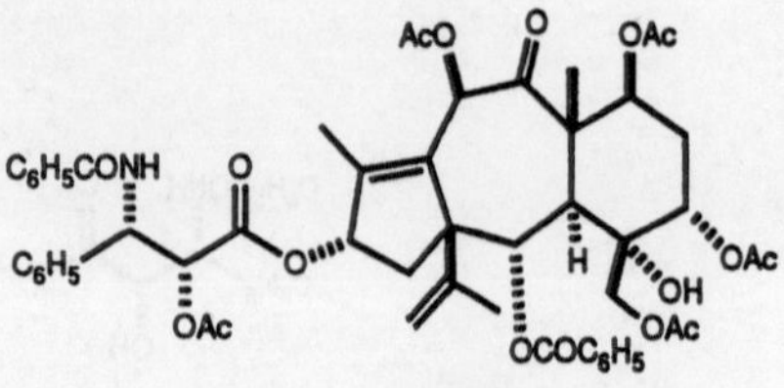

4.86

Analogous rearrangements of taxol have also been described briefly by POTIER and his collaborators resulting from the treatment of taxol with zinc chloride under fairly vigorous conditions (*87*). Although the structures proposed by these investigators involved methyl migration rather than ring contraction, it is possible that the products are in fact exactly analogous to the ring-contracted structure (**4.86**).

A taxol analog with a contracted A-ring but an intact oxetane ring was prepared by protection of the 2' and 7 hydroxyl groups as their triethylsilyl derivatives (**4.87**) and treatment with mesyl chloride and triethylamine. The 1-mesyl derivative (**4.88**) that is presumably formed under these conditions could not be isolated, but instead rearranged to the ring-contracted taxol derivative (**4.89**), which was deprotected to the A-nortaxol (**4.90**) (Scheme 4.13) (*230*). This taxol derivative retains much of the activity of taxol in the tubulin-assembly assay, although it is not especially active in the KB cell culture assay.

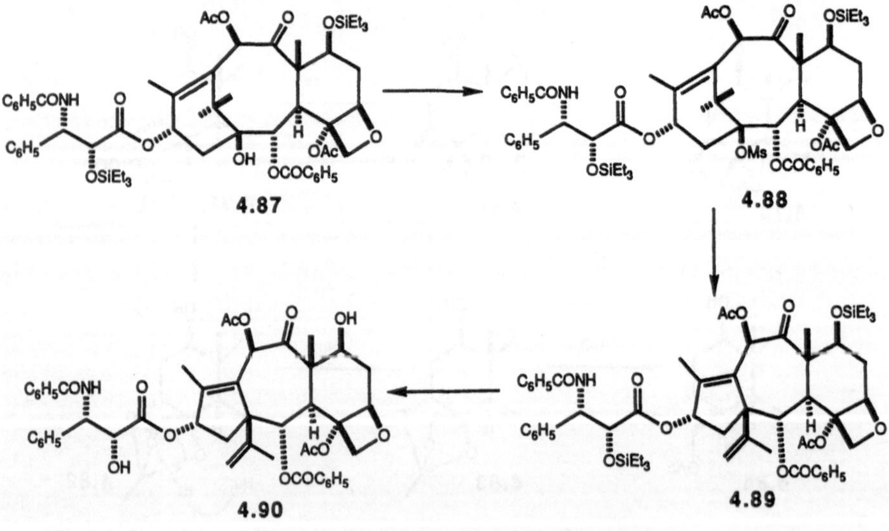

4.87

4.88

4.90

4.89

Scheme 4.13. Synthesis of A-nortaxol **4.90**

A similar ring contraction was observed by HALSALL and his collaborators in their studies of a taxoid related to taxusin. In this case, however, since their taxoid lacked a C-1 hydroxyl group, rearrangement was initiated by protonation of the C-13 hydroxyl group, leading to the formation of a product with the isopropenyl group at C-11 rather than at C-1. Thus treatment of the taxoid (4.91) with periodic acid in acetone gave the rearranged acetonide (4.13) (Scheme 4.14). The structure of (4.13) was established unambiguously by an X-ray crystal structure of its dihydro-*p*-bromobenzoate derivative (*32*).

Although taxoids such as taxol are unstable in base and give complex mixtures of products, an interesting rearrangement of taxinine has been observed in base. Treatment of taxinine or its 9,10-desacetyl derivative (4.92) with ethanolic potassium hydroxide leads to the formation of anhydrotaxininol (4.95); the mechanism proposed for this rearrangement is shown in Scheme 4.15. The formation of (4.95) was first observed as

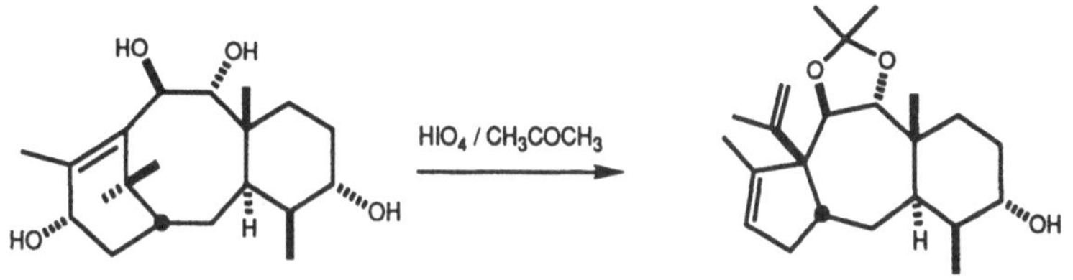

Scheme 4.14. Synthesis of A-nortaxoid 4.13

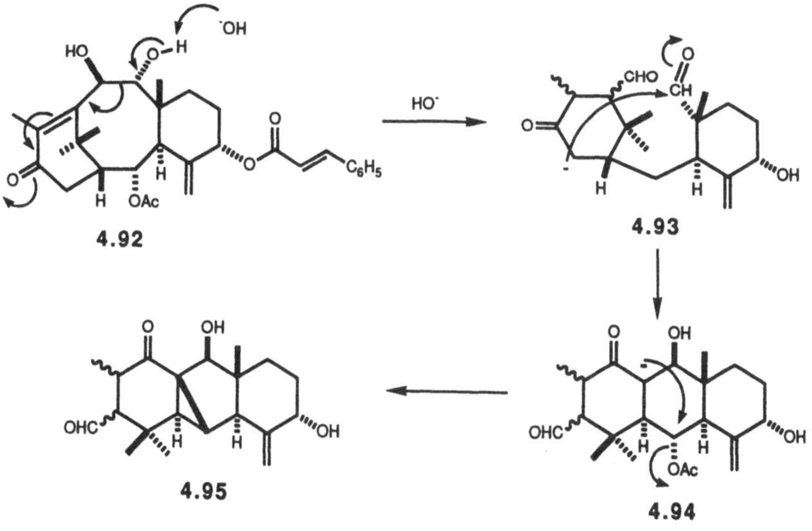

Scheme 4.15. Synthesis of anhydrotaxininol 4.95

early as 1931 by TAKAHASHI (*266*), but its correct composition was not established until 1960 (*264*) and its structure was finally established in 1964 (*265, 305*). Its formation was also used to help establish the stereochemistry of taxinine (*152*).

4.3.7. *Photochemistry*

The photochemistry of various taxinine derivatives has been studied primarily by NAKANISHI and his collaborators (*39, 143*). Two major types of reactions were observed. The first one is a transannular reaction between C-3 and C-11, leading to a tetracyclic product. Thus irradiation of the diacetate (**4.96**) in dioxane with a 450 W high-pressure mercury lamp with a Pyrex filter for 2 hours gave the transannular product (**4.97**) in quantitative yield (Scheme 4.16). Diacetylation of (**4.97**) afforded taxinine L, a minor component of *T. cuspidata* (*39*). The rearrangement, which formally involves a hydrogen transfer from C-3 to C-12 and bond formation between C-3 and C-11, was shown to be a triplet reaction, as it was sensitized with acetophenone and quenched with piperylene and cyclohexa-1,3-diene. A concerted $\sigma_s^2 + \pi_s^2$ route was proposed.

The photochemistry of taxinine takes a different course in the isopropylidene derivative (**4.98**). In this compound the C-3 and C-11 carbons are moved apart somewhat by the constraint of the additional ring system and irradiation of (**4.98**) gave the cyclopropyl ketone (**4.99**) as the major product (*143*). A similar result was obtained on irradiation of the dihydro derivative (**4.100**). In this compound the reduction of the exocyclic double bond converts the allylic H-3 into a simple tertiary hydrogen with correspondingly reduced reactivity. This reactivity change is enough to eliminate the transannular reaction even in the absence of any steric constraints, and a cyclopropyl ketone corresponding to (**4.99**) and the solvent adduct (**4.101**) are the major products of irradiation (*143*).

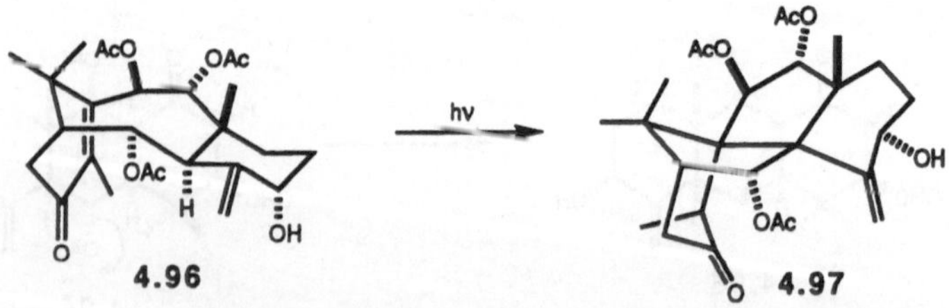

Scheme 4.16. Photochemistry of taxinine **4.96**

4.98

4.99

4.100

4.101

5. Approaches to the Synthesis of Taxane Diterpenoids

Although taxane diterpenoids have been known since 1856 (*163*) and the first structure determination was reported in 1963, the field of taxoid synthesis did not begin to develop until the early 1980's. Much of the effort toward the total synthesis of this class of compounds has been motivated by taxol (**5.1**), the most functionally and stereochemically complex taxoid. Thus GREENE and his collaborators, writing in 1988, state that "taxol is quite possibly the number one target today of synthetic organic chemists" (*55*). The great interest in taxol is a result of its potent antitumor and antileukemic properties as well as its unique mode of action. This, combined with the fact that taxol is isolated from the bark of the pacific yew tree in very low yield (approximately 100 mg per kilo of bark) (*19, 131*) by a method which is fatal to the tree, has prompted a "world wide .. prodigious effort towards its total synthesis." (*55*)

From a synthetic point of view, the abundance of stereochemical detail and high level of structural complexity make the taxoids one of the most synthetically challenging classes of compounds. A quick examination of the taxane diterpenoids will reveal that, with only a few exceptions, they are a group of compounds that share the tricyclic carbon skeleton (**5.2**). The degree and location of oxygenation on the skeleton can vary greatly. Although the complex functionality and stereochemical issues must ultimately be dealt with, the development of a general and

5.1   5.2

efficient method for the construction of a suitably substituted tricyclic carbon skeleton has often been recognized as the major task (199, 227, 250).

Despite the effort of over 30 research groups, only modest success in the total synthesis of naturally occuring taxoids has occurred. Although a large number of synthetic strategies have been published, the synthesis of (−)-taxusin (the unnatural enantiomer) (5.3) by HOLTON et al. (109) remains the only report of the synthesis of a natural taxoid. The focus of most research groups has been the preparation of partial structures in an effort to explore various concepts that deal with the construction of the strained tricyclo[9.3.1.0³·⁸]pentadecane skeleton equipped with functionalization suitable for further elaboration to various taxoids.

5.3

This part of the review focuses on the synthetic efforts toward the tricyclic diterpenoid structure of the taxoids. Since two very comprehensive reviews (207, 255) as well as at least three partial reviews (17, 153, 203) have appeared recently on this subject, the intent here is to give only a brief overview for the reader. For an in depth discussion of the subject, the reader is referred to the previously mentioned reviews. Since taxol has been the stated target of many groups, this section will also include a review of synthetic strategies for the preparation and attachment to baccatin III of the taxol A-ring side chain and for the synthesis of the 3-oxygenated oxetane ring. The classification scheme used here is similar to that used by SWINDELL (255).

5.1. Linear Strategies

5.1.1. Biomimetic Approaches

One of the earliest biogenetic proposals for the taxoids was published by LYTHGOE in 1966 (*96*). LYTHGOE suggested that the taxoids are formed from geranylgeranyl pyrophosphate by electrophilic cyclization (**5.4** → **5.6**), possibly via cembrene (**5.7**) or verticillene (**5.8**) intermediates (Scheme 5.1). Although this is the most commonly accepted hypothesis, there is no experimental support for this proposal. Several groups have investigated approaches modeled after LYTHGOE's biogenetic hypothesis, but the taxoid framework still remains to be prepared by this strategy.

5.1.1.1. Kato's Approach

KATO and coworkers have reported the synthesis of two *seco*-taxoid derivatives using a strategy based on LYTHGOE's biogenetic proposal. The actual route employed in the construction of these *seco*-taxoid derivatives was a slight modification of the biogenetic hypothesis since the C9-C10 bond was connected in the last stage of the synthesis. KATO's initial report described the synthesis of *seco*-taxoid derivative (**5.12**) (*128*) (Scheme 5.2). The key cyclization intermediate (**5.11**) was prepared by coupling geranyl cyanide (**5.9**) and allyl chloride (**5.10**). Cyclization of key intermediate (**5.11**) generated (**5.12**).

In 1981, KATO *et al.* described an attempt to prepare *endo*-anhydro-verticillol (**5.16**) using the same strategy (*149*) (Scheme 5.3). In this case,

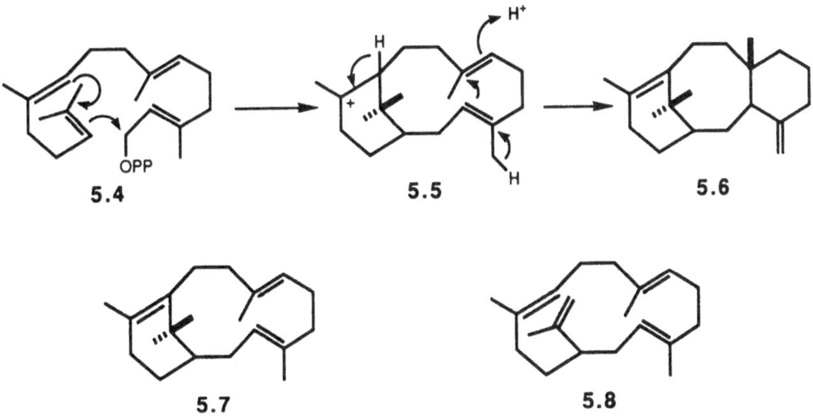

Scheme 5.1. Lythgoe's biogenetic hypothesis for the taxoid skeleton

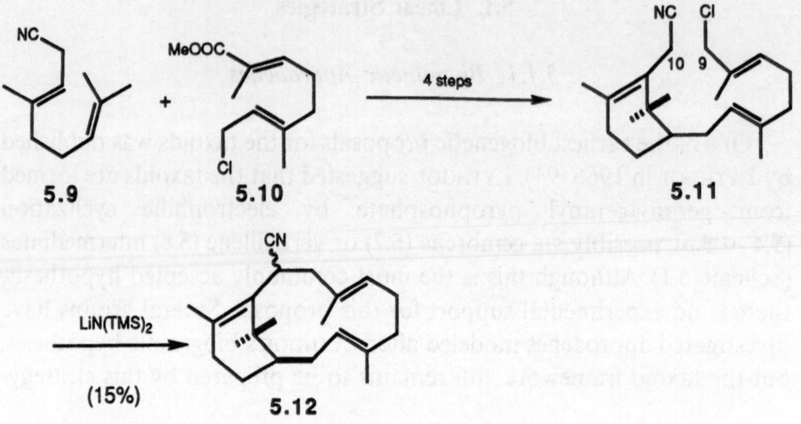

Scheme 5.2. Kato's first approach for the synthesis of the seco-taxoid skeleton

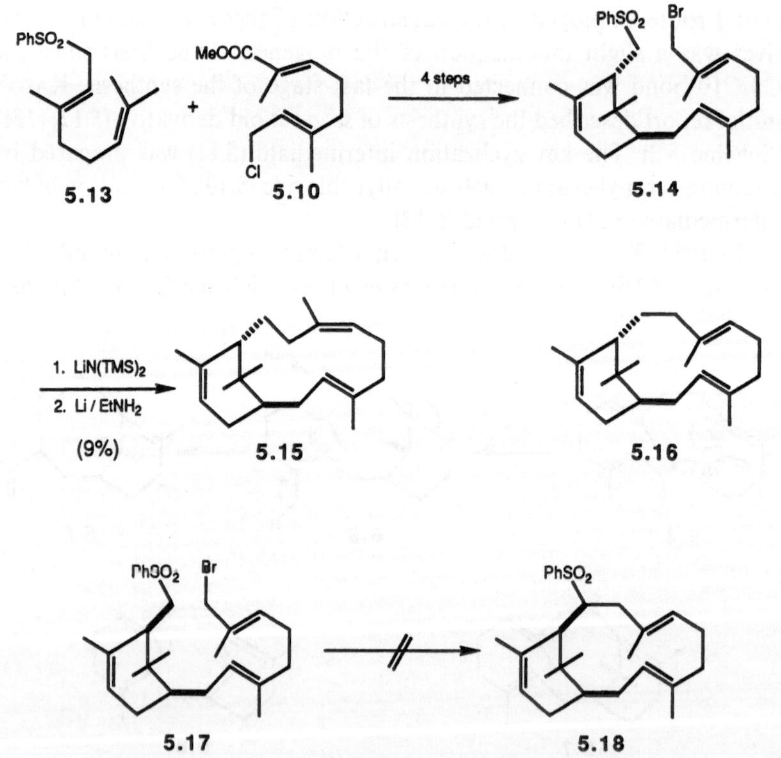

Scheme 5.3. Kato's second synthesis of the seco-taxoid skeleton

the key intermediate (**5.14**) was prepared by coupling neryl phenyl sulfone (**5.13**) with allyl chloride (**5.10**). Unfortunately, cyclization of (**5.14**) produced (**5.15**), the 7Z-isomer of anhydroverticillol. Several attempts were also made to cyclize (**5.17**), but these failed to produce the desired product, (**5.18**).

5.1.1.2. Frejd's Approach

FREJD and coworkers (*77, 211*) reported the enantiospecific synthesis of a taxol A-ring derivative that possesses functionality suitable for further elaboration of the BC rings of the taxoid skeleton. The optically active A-ring segment (**5.23**) was synthesized from L-arabinose (**5.18**) in 23 steps (Scheme 5.4). The synthetic scheme was based on the cyclization of the epoxy-allylsilane (**5.21**), a process modeled after LYTHGOE's biogenetic proposal for the taxoid A-ring. The key cyclization intermediate (**5.21**) was formed from L-arabinose (**5.18**) using a route that passed

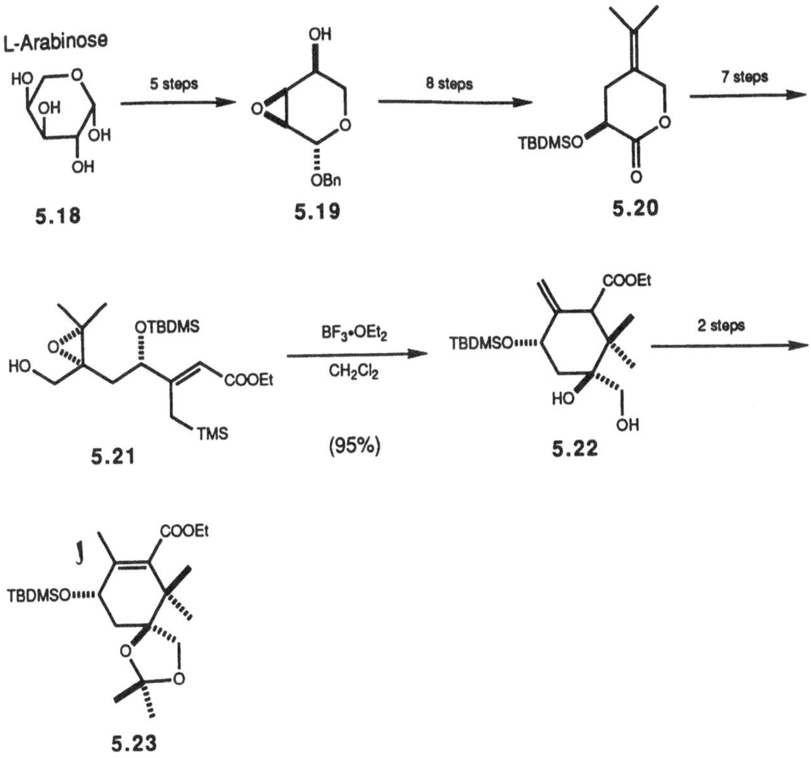

Scheme 5.4. Frejd's enantiospecific synthesis of the taxoid A-ring

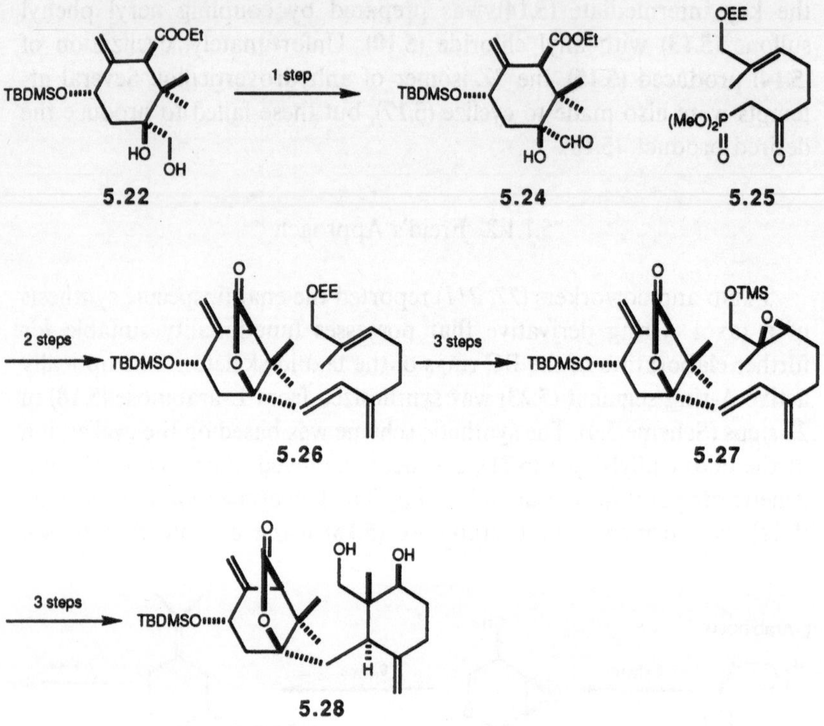

Scheme 5.5. Frejd's synthesis of a 9,10-seco-taxoid derivative

through intermediates (5.19) and (5.20). Cyclization of epoxy-allylsilane (5.21) led to the highly functionalized cyclohexane (5.22) which was then converted to A-ring precursor (5.23) in two steps. This cyclization provided a unique example of a "polyene-type" cyclization of an epoxy-olefin where the epoxide is tetrasubstituted.

The most recent disclosure (76) by FREJD describes the extension of this work to include the preparation of 9,10-seco-taxoid (5.28) (Scheme 5.5). This sequence began with diol (5.22), an intermediate available from his A-ring synthesis. After oxidation of (5.22) to give aldehyde (5.24), application of Witting chemistry joined C-ring precursor (5.25) to (5.24) to give tetraene (5.26). A three step procedure then converted (5.26) to epoxide (5.27). The final series of steps led to 9,10-secotaxoid (5.28).

5.1.1.3. Pattenden's Approach

PATTENDEN et al. reported the synthesis of E,E-verticillene (5.7) (Scheme 5.6) (8, 119) with the intention of using it to test the latter part of

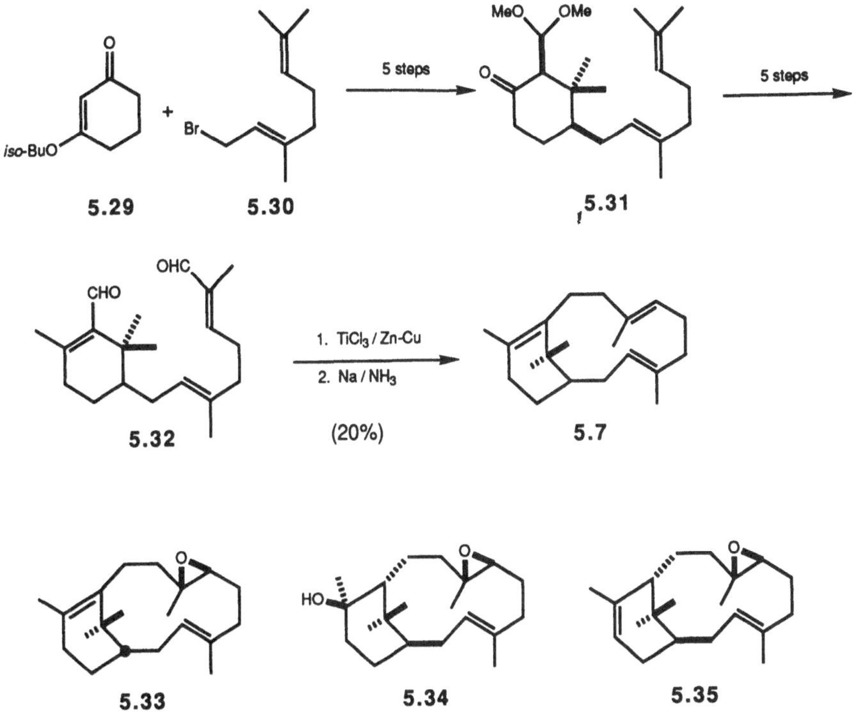

Scheme 5.6. Pattenden's synthesis of verticillene

LYTHGOE's hypothesis, which suggests that the taxoids are formed in nature via verticillene-type intermediates by transannular cyclization to form the C-ring. Like the compounds prepared by KATO, verticillene possesses the *seco*-taxoid skeleton. PATTENDEN's synthesis was based on the intramolecular reductive coupling of bis-aldehyde (**5.32**) in the presence of Ti(O), with concomitant 1,5-H sigmatropic rearrangement to generate the twelve-membered ring followed by 1,4 reduction of the resulting tetraene intermediate to give E,E-verticillene. This key intermediate was prepared from 3-isobutoxycyclohex-2-enone (**5.29**) and geranyl bromide (**5.30**) via intermediate (**5.31**).

Once PATTENDEN had verticillene in hand, he proceeded to investigate the possibility of transannular cyclization using Lewis acids to form the taxoid skeleton (*7, 8*). All attempts to cyclize verticillene (**5.7**) as well as the epoxide derivatives (**5.33**)–(**5.35**) failed to produce the desired tricyclic taxoid ring system; these cyclizations are discussed in more detail in Sect. **6.1**.

5.1.2. Intramolecular Diels-Alder Approaches

The intramolecular Diels-Alder reaction has often played an important role in the synthesis of complex polycyclic molecules. The trimethyl substituted six-membered A-ring and a *trans*-fused six-membered C-ring present in the parent taxoid ring system (**5.2**) have prompted the investigation of several intramolecular Diels-Alder approaches in the synthesis of taxoids. Both the A-ring and the C-ring have been targets.

5.1.2.1. Shea's Approach

Early work by SHEA and coworkers revealed that bridgehead alkenes could be formed via the intramolecular Diels-Alder reaction under either thermal (*244*) or Lewis acid catalyzed (*239*) conditions when the dienophile is joined to the diene at the 2 position. This prompted SHEA to investigate the possibility of using this strategy in the synthesis of the taxoid skeleton.

In 1983, SHEA *et al.* reported the synthesis of the aromatic taxoid derivatives (**5.43**) and (**5.44**) (*237*, *238*). The key cycloaddition substrates (**5.41**) and (**5.42**) were prepared by coupling either (**5.36**) or (**5.37**) with benzyl Grignard (**5.38**) to give the corresponding dienes (**5.39**) and (**5.40**) which were then elaborated to (**5.41**) and (**5.42**) (Scheme 5.7). Cyclization of (**5.41**) and (**5.42**) under thermal conditions led to (**5.43**) and (**5.44**). SHEA

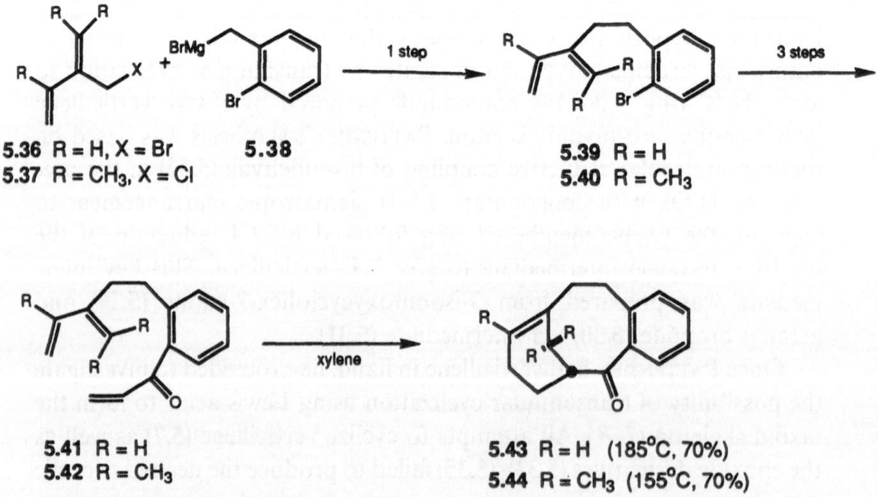
R
R
R
X
BrMg
+
Br
1 step
R
R
R
Br
3 steps
5.36 R = H, X = Br
5.37 R = CH₃, X = Cl
5.38
5.39 R = H
5.40 R = CH₃
R
R
R
O
xylene
R
R
R
O
5.41 R = H
5.42 R = CH₃
5.43 R = H (185°C, 70%)
5.44 R = CH₃ (155°C, 70%)

Scheme 5.7. Shea's synthesis of the C-aromatic taxoid skeleton

has also reported the synthesis of **(5.43)** by Lewis acid catalyzed cycliz-ation (*239*).

Once SHEA had synthesized **(5.43)** and **(5.44)**, he took advantage of the opportunity to examine the relationship between transition-state con-formation and the products produced in the Lewis acid-catalyzed intra-molecular Diels-Alder reaction that led to these products (*240*). Force-field calculations, molecular modeling and NMR methods were used to establish the parallel between transition-state and product stabilities.

SHEA expanded the scope of this investigation by synthesizing the methoxy trienone **(5.49)** (Scheme 5.8) (*241*). The construction of **(5.49)** began by coupling **(5.45)** with **(5.46)** to give dienol **(5.47)**. A three step sequence then elaborated **(5.47)** to the key cycloaddition substrate **(5.48)**. Cyclization of **(5.48)** under thermal or Lewis acid-catalyzed conditions led to **(5.49)**, which was a mixture of the two atropisomers, *endo*-**(5.49)** and *exo*-**(5.49)**. Again, taking advantage of this opportunity he investig-ated the conformational properties of these two products.

In 1988, SHEA and HAFFNER disclosed (*242*) the results of an investiga-tion aimed at applying his intramolecular Diels-Alder strategy to the

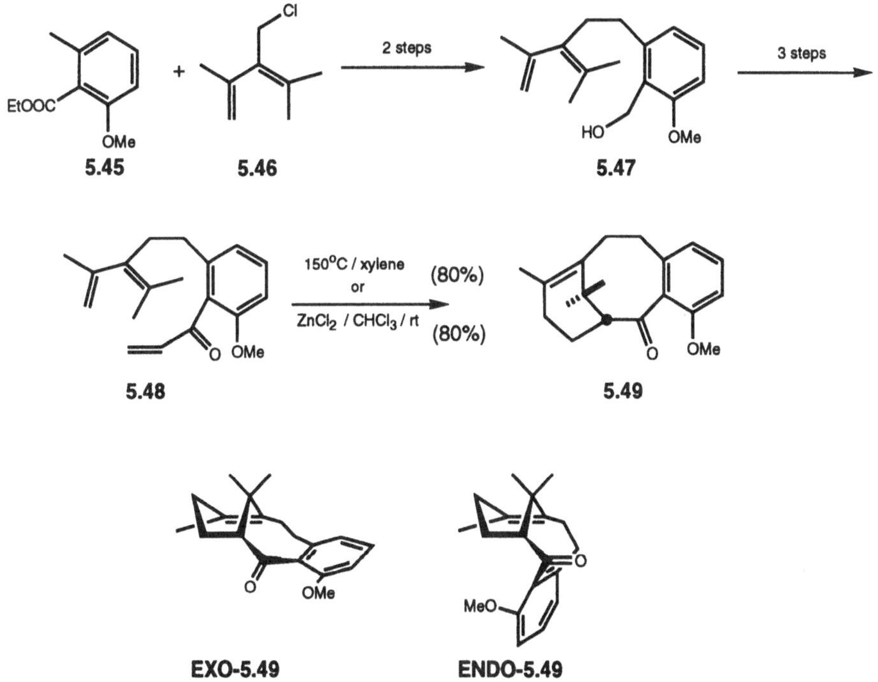

Scheme 5.8. Shea's synthesis of the C-7 methoxy C-aromatic taxoid derivative

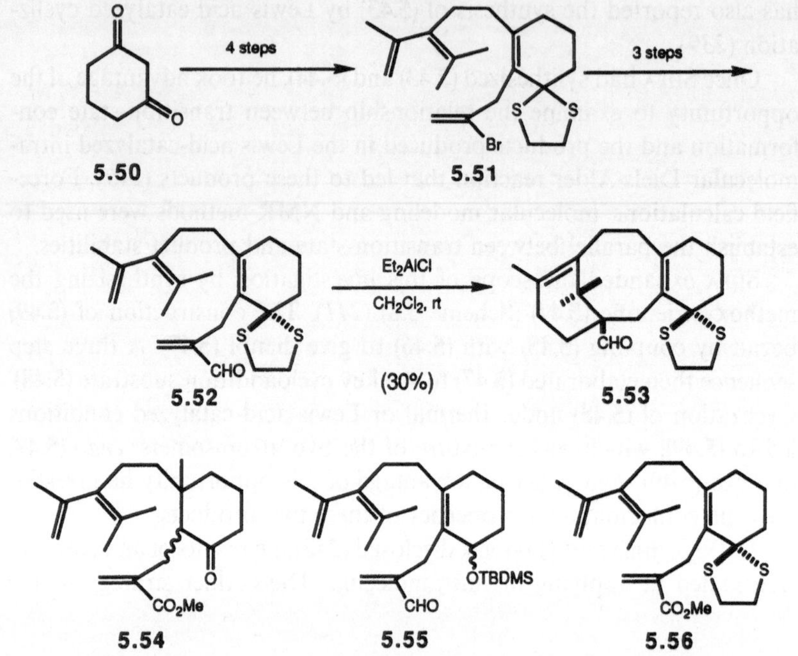

Scheme 5.9. Shea's synthesis of a partially saturated C-ring taxoid model

construction of a tricyclic taxoid model with a partially saturated C-ring that would allow the incorporation of functionality at C-1 (Scheme 5.9). This investigation began with the five step conversion of diketone (**5.50**) to bromodiene (**5.51**). A three-step sequence then produced key intermediate (**5.52**). Under either thermal or Lewis acid catalyzed conditions cyclization of (**5.52**) occurred to give the tricyclic model (**5.53**) but yields were disappointingly low in both cases. Attempts to cyclize (**5.54**)–(**5.56**) proved to be even less effective.

The most recent publication from this group (*243*) describes the results of a study designed to explore the diastereoselectivity with the steric size substituents at C-4 (of hydride reductions of the C-2 carbonyl). Structures analogous to (**5.49**) were used and both electrophilic and nucleophilic hydride reagents were examined.

5.1.2.2. Jenkins' Approach

JENKINS' strategy for the preparation of the tricyclo[9.3.1.0^{3,8}]-pentadecane ring system, like SHEA's, was based on an intramolecular

Diels-Alder cyclization of a 2-substituted diene fragment to construct both the A and B rings. His initial target was the tricyclic model system (5.60) (22, 23) (Scheme 5.10). Although this system lacked the three A-ring methyl groups present in naturally occurring taxoids, it possessed a saturated C-ring with the requisite stereochemistry at C-3 and C-8. The key cycloaddition substrate, (5.59), was prepared in eleven steps from enol ether (5.57) via intermediate (5.58). Lewis acid catalyzed Diels-Alder cyclization of (5.59) produced (5.60) in 72% yield.

In 1987, JENKINS et al. expanded this work to include the preparation of the fully methylated taxoid model (5.62) (20, 21) (Scheme 5.11). The construction of (5.62) began with (5.58), an intermediate from his initial synthesis. A six step sequence was used to convert (5.58) to (5.61) which was then cyclized to give the methylated taxoid model (5.62).

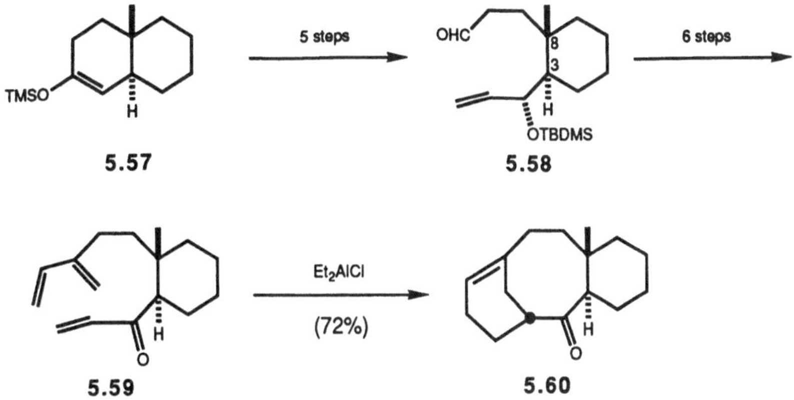

Scheme 5.10. Jenkins' first synthesis of the taxoid ring system

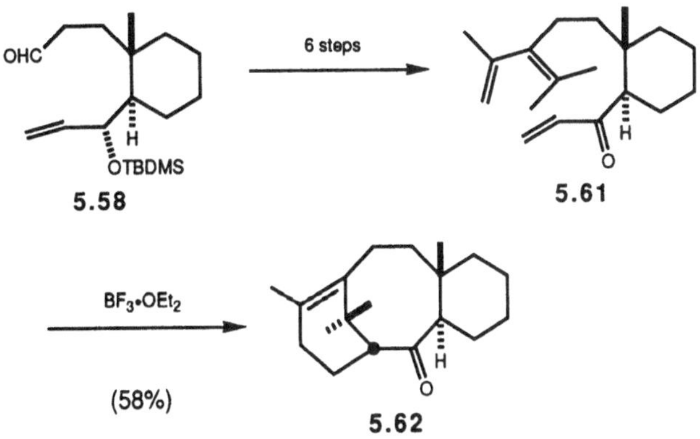

Scheme 5.11. Jenkins' synthesis of an alkylated taxoid skeleton

5.1.2.3. Sakan's Approach

SAKAN'S strategy for the construction of the taxoid skeleton was also based on an intramolecular Diels-Alder reaction, but unlike SHEA or JENKINS he focused his efforts toward the formation of the C-ring (227, 228) (Scheme 5.12). SAKAN chose to use the conformationally rigid bicyclo[2.2.2]octene system (5.65) in hopes of attaining maximum entropic assistance for the cyclization. This substrate had carbon atoms C-2, C-1, C-15, C-11, and C-10 rigidly positioned by the boat A-ring conformation so as to facilitate the closure he desired. This key intermediate was prepared from enone (5.63) via (5.64). Cyclization under thermal conditions resulted in the taxoid model (5.66). SAKAN also investigated the Lewis acid catalyzed intramolecular Diels-Alder reaction although the reaction proceeded in 90% yield the dominant product was the undesired C-8 epimer.

5.1.2.4. Yadav's Approach

A recent report (304) by YADAV and RAVISHANKAR describes a strategy for the construction of the taxoid skeleton centered around methodology employing an intramolecular Diels-Alder reaction followed by a Wittig rearrangement. This approach led to the tricyclic taxoid model (5.72) in 8 steps (Scheme 5.13). The Diels-Alder substrate (5.69) was prepared from bromodiene (5.67) and diol (5.68) in four steps. Lewis acid catalyzed cyclization of (5.69) led to (5.70) which was then converted to rearrangement substrate (5.71). Treatment with n-BuLi induced a Wittig rearrangement to give tricycle (5.72).

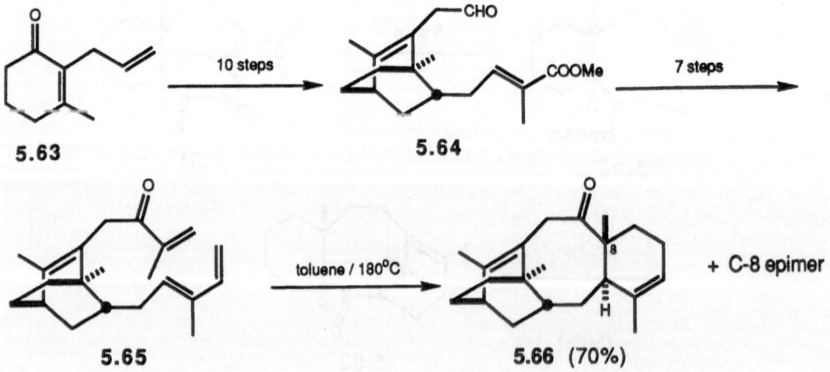

Scheme 5.12. Sakan's approach to taxoid system

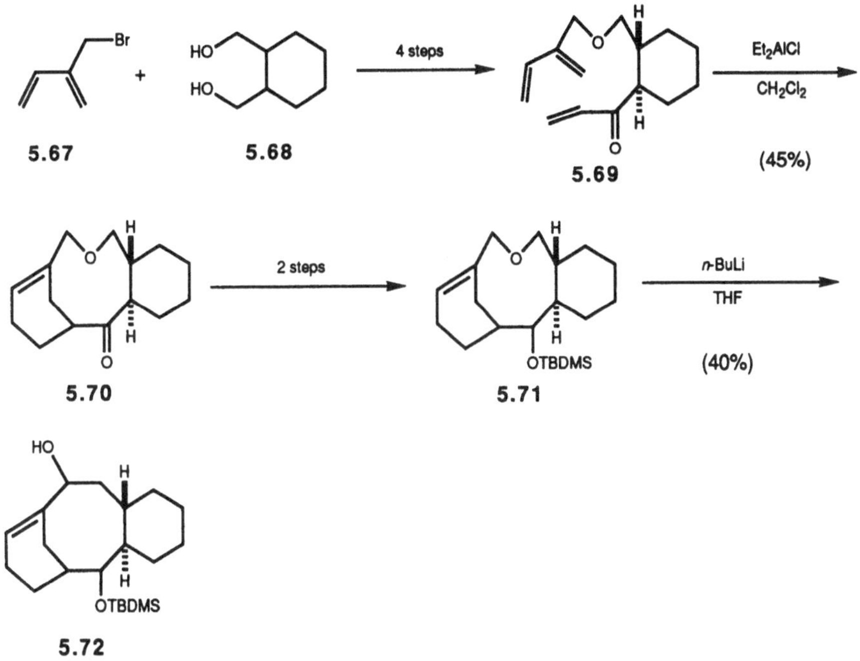

Scheme 5.13. Yadav's synthesis of a tricyclic taxoid model

5.1.3. AB → ABC Approaches

The majority of linear approaches fit into this category.

5.1.3.1. Martin's Approach

The efforts of MARTIN and coworkers were focused on the construction of a series of substituted bicyclo[5.3.1]undec-7-enes which could be considered AB intermediates (173). The general three-step sequence reported involved the skeletal reorganization of bicyclo[2.2.2]octadienol (5.74), prepared in two steps from diketone (5.73), via an anionic oxy-Cope rearrangement to yield compounds of the general structure (5.76) (Scheme 5.14). Since the key intermediate dienol was produced as a diastereomeric mixture of alcohols (5.74) and (5.75), MARTIN et al. examined a variety of conditions in an attempt to improve the diasteroselectivity. Unfortunately, this effort went unrewarded. Using this methodology, they prepared model compounds (5.77)–(5.81).

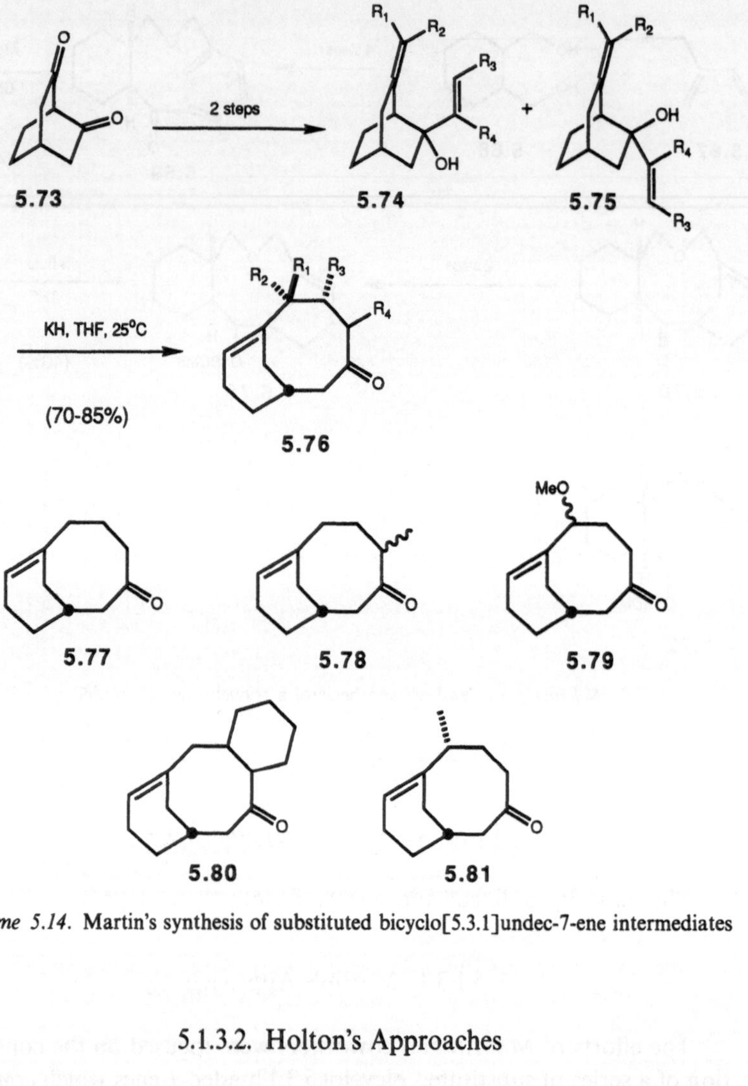

5.73 5.74 5.75 5.76 5.77 5.78 5.79 5.80 5.81

Scheme 5.14. Martin's synthesis of substituted bicyclo[5.3.1]undec-7-ene intermediates

5.1.3.2. Holton's Approaches

HOLTON's initial efforts were directed toward the synthesis of the optically active tricyclic taxane ring system (**5.85**) (*106*) (Scheme 5.15). This taxane model was prepared from (–)-β-patchoulene oxide (**5.82**) in five steps with an overall yield of 53%. The strategy involved the fragmentation of (**5.83**) followed by annulation of ring C onto the resulting hydroxy ketone (**5.84**).

An extension of this approach led to the first synthesis of a naturally occurring taxoid, an achievement which was the highlight of synthetic taxoid work in the 1980's. The compound prepared was taxusin (**5.3**), or

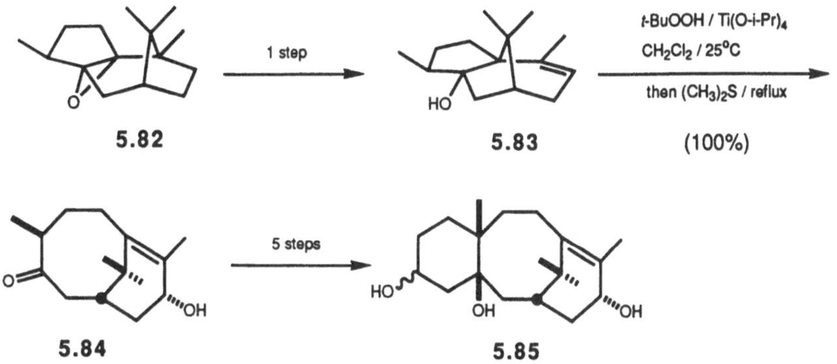

5.82 → (1 step) → **5.83**

t-BuOOH / Ti(O-i-Pr)₄
CH₂Cl₂ / 25°C
then (CH₃)₂S / reflux

(100%)

5.84 → (5 steps) → **5.85**

Scheme 5.15. Holton's synthesis of an AB taxoid model

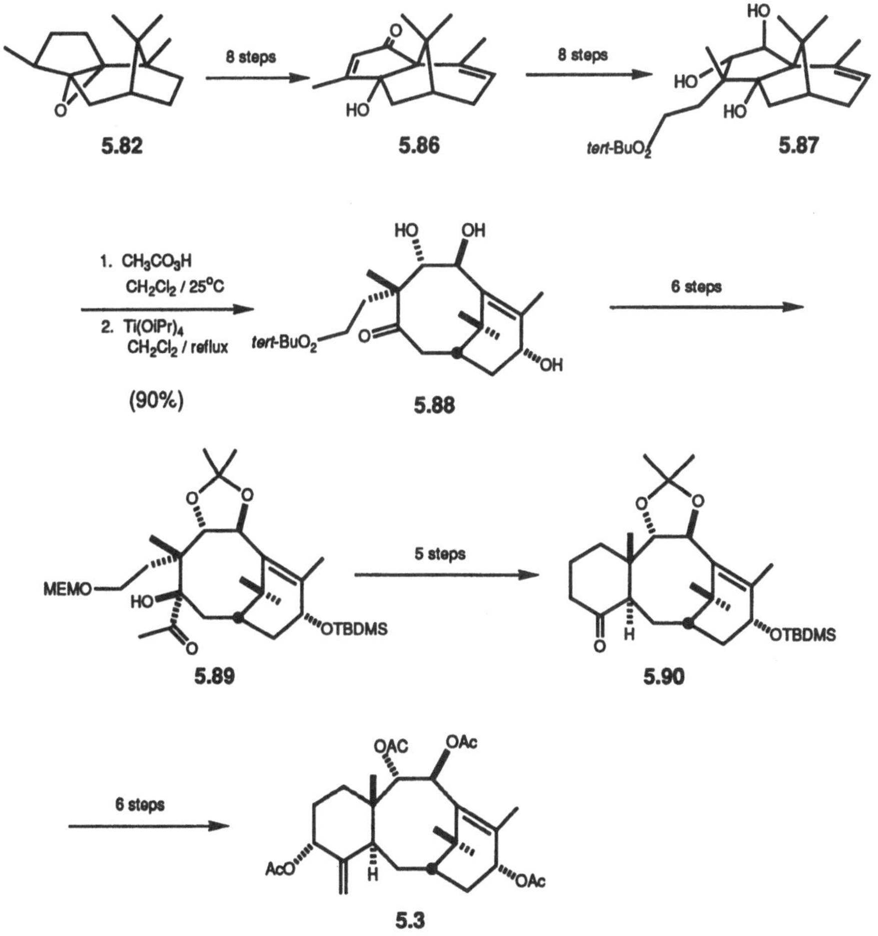

5.82 → (8 steps) → **5.86** → (8 steps) → **5.87**

tert-BuO₂

1. CH₃CO₃H
   CH₂Cl₂ / 25°C
2. Ti(OiPr)₄
   CH₂Cl₂ / reflux   tert-BuO₂

(90%)   **5.88** → (6 steps) →

5.89 → (5 steps) → **5.90**

MEMO

OTBDMS   OTBDMS

(6 steps) → **5.3**

OAc   OAc

AcO

OAc

Scheme 5.16. Holton's synthesis of (−)-taxusin

more correctly *ent*-taxusin, since the product was the enantiomer of the natural product (Scheme 5.16) (*109*). As with the preparation of the model system, HOLTON *et al.* started with (−)-β-patchoulene oxide (**5.82**). In eight steps he generated enone (**5.86**), which was then elaborated to the key intermediate (**5.87**). Following epoxidation of (**5.87**), fragmentation was induced using Ti(OiPr)$_4$ to form the B-ring. This produced triol (**5.88**) in 90% yield from (**5.87**). The remainder of the synthesis focused on the elaboration of the C-ring. In order to achieve this, intermediate (**5.89**) was prepared first to introduce the last two carbons of the C-ring. Finally, a five step sequence was used to complete the C-ring and a further six steps completed the synthesis of taxusin. When performed in the most efficient way, (−)-taxusin was prepared in > 20% overall yield.

It is unfortunate that the most readily available enantiomer of the starting material, (−)β-patchoulene oxide, leads to the *ent*-taxoid skeleton. If (+)-β-patchoulene oxide, which can be prepared from (−)-camphor following BÜCHI's procedure (*25*) were to be used, taxoids of the correct absolute stereochemistry could be prepared, but this conversion would add several steps to the synthesis. In addition, because of the constraints of working with a naturally occurring starting material, this approach would require considerable modification for it to be developed into a synthesis of baccatin III.

5.1.3.3. Oishi's Approach

In the early 1980's, OISHI and OHTSUKA developed a general method for the construction of medium ring ketones based on the contraction of lactam sulfoxides or sulfones through transannular acylation of a sulfur-stabilized anion (*193, 194, 195*). Using this new methodology, they prepared the AB-ring taxoid model (**5.94**) (*196, 199*) (Scheme 5.17). This was accomplished by first preparing the A-ring precursor (**5.92**) from β-ionone (**5.91**). A seven step procedure then converted (**5.92**) to the key twelve-membered lactam-sulfide (**5.93**) which underwent ring contraction to yield the target compound (**5.94**). Although this provided a test of the ring forming methodology, the procedure developed made it difficult to introduce any functionality in the A-ring precursor. Therefore, this model was not considered a viable intermediate for further elaboration to the taxoid skeleton.

To remedy this situation, OISHI and OHTSUKA turned their attention to the development of a route that would allow the incorporation of a double bond in the A-ring precursor. This goal was achieved with the

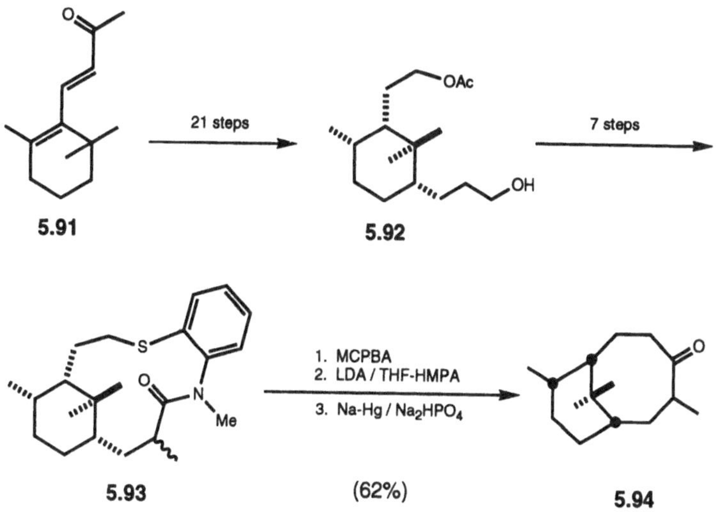

Scheme 5.17. Oishi's first synthesis of an AB taxoid model

preparation of taxoid models **(5.99)** and **(5.102)** (*197, 200*). The route that led to **(5.99)** (Scheme 5.18) began with **(5.95)**, which was prepared from α-ionone in two steps. The A-ring precursor **(5.97)** was synthesized from **(5.95)** via a route that passed through intermediate **(5.96)**. With this in hand, the medium ring forming methodology was employed to yield lactam-sulfide **(5.98)** which then underwent ring contraction to yield the

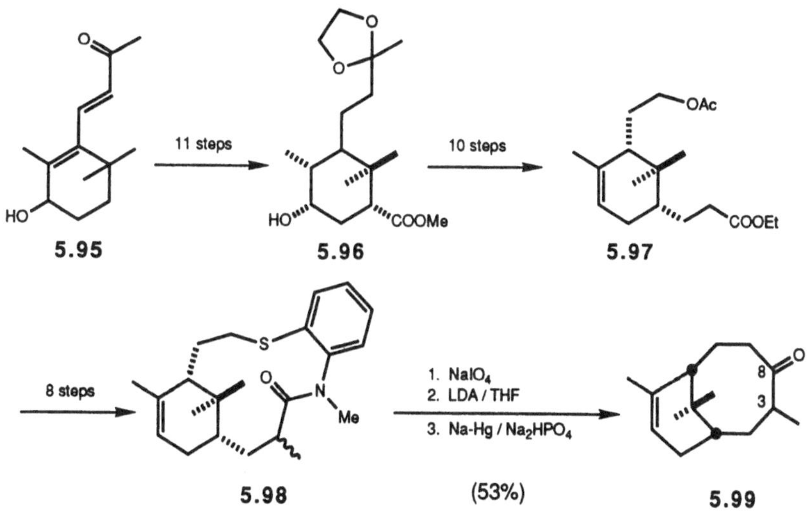

Scheme 5.18. Oishi's second synthesis of an AB taxoid model

bicyclic compound (5.99). Although this model had the desired A-ring double bond, it suffered from the fact that the methyl group was located at C-3 rather than at C-8 as found in the taxoid skeleton.

OISHI and OHTSUKA provided a solution to this problem with the synthesis of (5.102) (Scheme 5.19). Starting from the previously prepared intermediate (5.96), an eight step procedure yielded A-ring precursor (5.100). Then, using a slight modification of the previously used eight-membered ring forming sequence, the lactam sulfide (5.101) was prepared. Application of the ring contraction sequence gave the desired model compound, (5.102). Unfortunately the route to (5.102) was not very practical because it involved two reactions that proceeded very slowly, with one taking 36 days. Since (5.102) was considered a potential intermediate for the preparation of the tricyclo[9.3.1.0³,⁸]undecane ring system, OHTSUKA and OISHI have developed a more efficient strategy for its preparation (202).

Although complete experimental details are not available, OHTSUKA and OISHI have also reported the preparation of the tricyclic taxoid

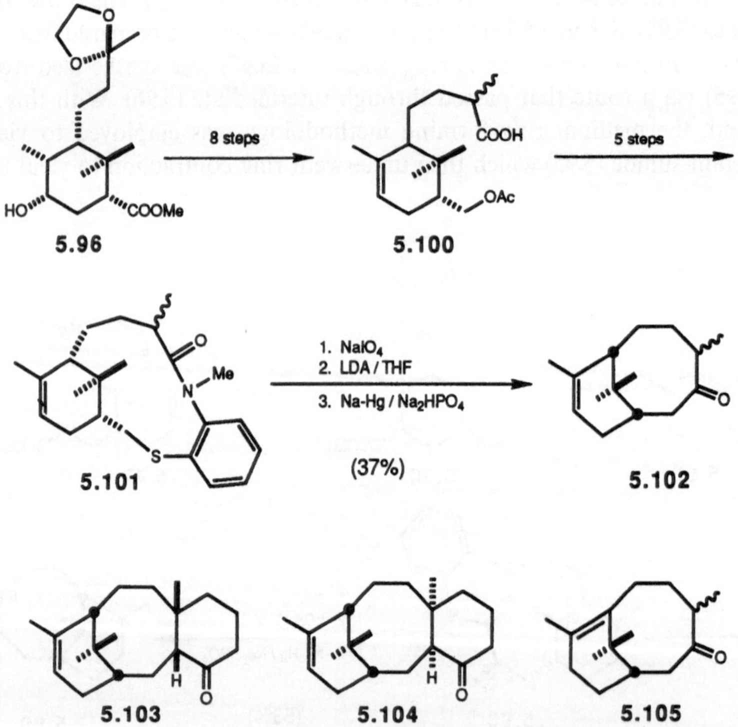

Scheme 5.19. Oishi's modified approach to bicyclic and tricyclic models

models (**5.103**) and (**5.104**) (*198*) as well as the synthesis of the AB
intermediate (**5.105**) (*201*).

5.1.3.4. Fétizon's Second Approach

In light of the difficulty he encountered with the construction of the
B-ring by direct cyclization (see 5.2.1.2), FÉTIZON and coworkers turned
their attention to the preparation of the taxoid skeleton using a [2 + 2]
photocycloaddition followed by a retroaldol cleavage of the resulting
cyclobutanone ring (de Mayo reaction). Initially, he investigated a series
of simple model compounds in an effort to determine the feasibility of
this approach (*9*). These studies resulted in the synthesis of model
compounds (**5.106**) and (**5.107**).

5.106   **5.107**

Using the information gleaned from the model studies, they used this
methodology to construct the AB-ring taxoid model (**5.111**) (Scheme
5.20) (*10, 74*). Starting with dione (**5.108**), key intermediate (**5.109**) was
prepared in eight steps. When this intermediate, in the form of its enol

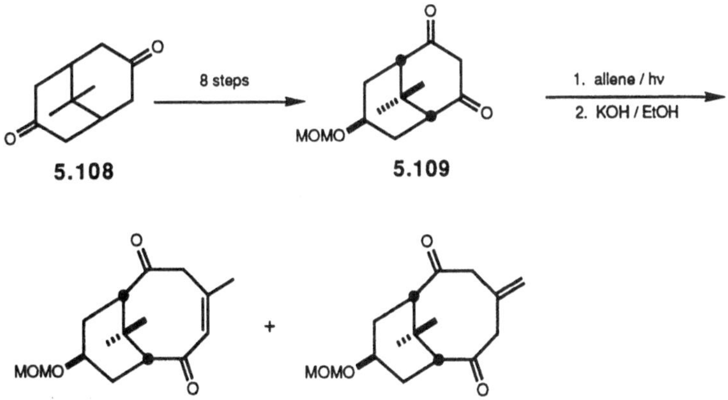

Scheme 5.20. Fétizon's photochemical approach to AB taxoid models

tautomer, was subjected to the [2 + 2] photocycloaddition-fragmentation sequence, double bond regioisomers (**5.110**) and (**5.111**) were produced in 73% combined yield from (**5.109**).

5.1.3.5. Blechert's Second Approach

Like FÉTIZON, BLECHERT's group has pursued the synthesis of the
taxoid skeleton using a [2 + 2] photocycloaddition followed by fragmentation of the resulting cyclobutane. This approach was a modification of the methodology developed in earlier work (*189, 190*) (see 5.2.2.3),
and was designed to overcome the difficulties they had encountered in
trying to install the C-19 methyl group.

The second approach by BLECHERT and coworkers centered on the
preparation of the AB-ring taxoid intermediate (**5.115**) (*124*) (Scheme
5.21). This bicyclic intermediate is structurally similar to FÉTIZON's AB
models, (**5.110**) and (**5.111**). The synthesis started with the transformation
of (**5.112**) to key intermediate (**5.113**) using an eight step sequence.
Photoaddition of this intermediate to allene produced (**5.114**) in 83%
yield. BLECHERT also attempted this same photoaddition using the benzyl
carbonate and trichloroethyl carbonate analogues of (**5.113**), but both
proved to be less effective. Finally, (**5.114**) was subjected to palladium-
mediated deprotection followed by base-induced fragmentation to yield
the target molecule (**5.115**).

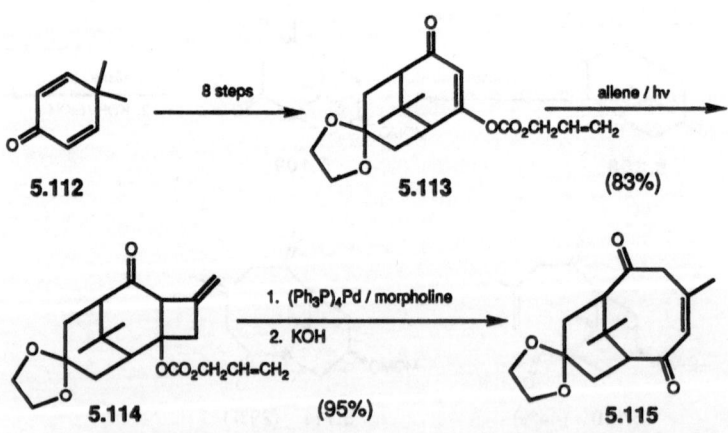

Scheme 5.21. Blechert's synthesis of an AB taxoid model

References, pp. 173–189

5.1.3.6. Wender's C-Ring Annulation Approach

WENDER'S strategy for the preparation of the taxoid skeleton is based on new methodology he developed for the construction of polycycles containing eight-membered rings using the nickel catalyzed [4 + 4] cycloaddition of tethered 1,3-dienes (287, 288). One of the unique features of this cycloaddition methodology is that it can be extended to the preparation of both AB and BC intermediates (290).

To demonstrate the applicability of this new methodology to taxoid systems, WENDER et al. prepared a model compound that resembled the AB portion of the taxoid skeleton (290) (Scheme 5.22). The key cycloaddition precursor (5.117) was prepared from commercially available myrcene (5.116) using a four step sequence that resulted in 93% E selectivity and a 27% overall yield. Cycloaddition of (5.117) using Ni(COD)₂ in the presence of triphenylphosphine produced (5.118) and (5.119) in a combined yield of 52% at 68% conversion.

The most recent report (289) from WENDER's group describes the preparation of a second AB model using a similar approach (Scheme 5.23). In this case, he transformed commercially available 4-pentyn-1-ol (5.120) into diene (5.121) in three steps. After conversion to the key tetraene (5.122), the AB model was formed by nickel catalyzed cyclization. When Ni(COD)₂ was used in the presence of tri-o-biphenyl phosphite, the result was a 7:1 mixture of (5.123) and (5.124) in a combined yield of 74%. In order to demonstrate the versatility of these cycloadducts, (5.123) was deprotected to give alcohol (5.125) and then selectively oxidized to epoxy ketone (5.126).

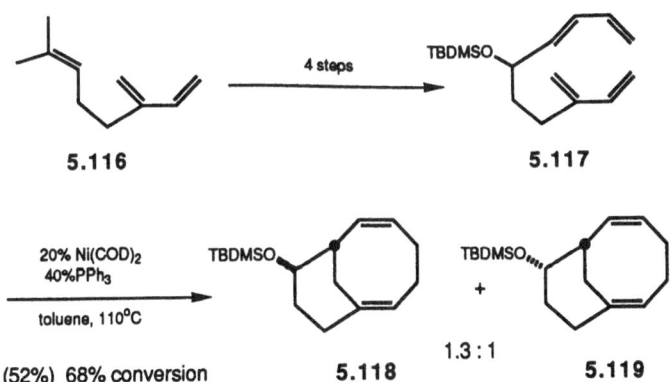

Scheme 5.22. Wender's initial synthesis of bicyclo[5.3.1]undecadienes

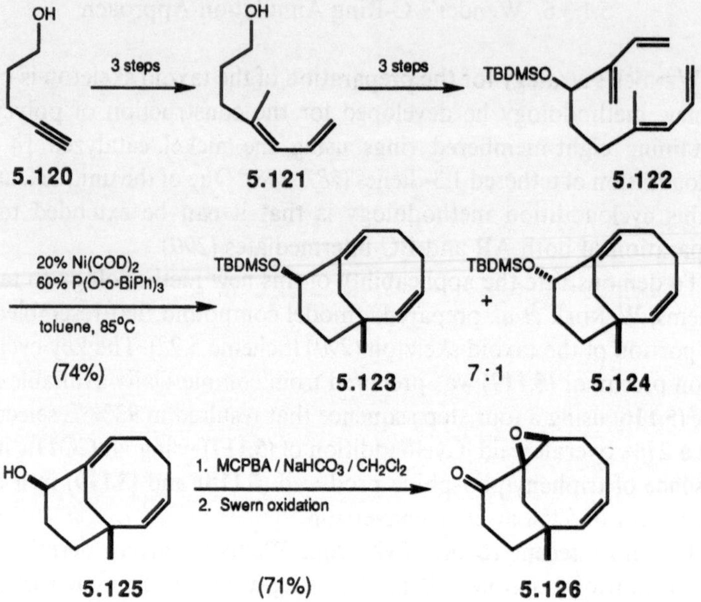

Scheme 5.23. Wender's most recent approach to bicyclo[5.3.1]undecadienes

5.1.3.7. Kraus' Approach

KRAUS and coworkers reported (147) a three-step protocol for the construction of bicyclo[5.3.1]undecene (5.129) (Scheme 5.24). This AB portion of the taxoid skeleton was constructed by a two carbon ring expansion of bicyclo[3.3.1]nonane (5.128) using a bridgehead intermediate that was generated *in situ* from the bromoketone (5.127).

5.1.3.8. Yamada's Approach

In 1984, YAMADA's group described (186) the preparation of the bicyclo[5.3.1]undecane derivative (5.133) using a strategy centered

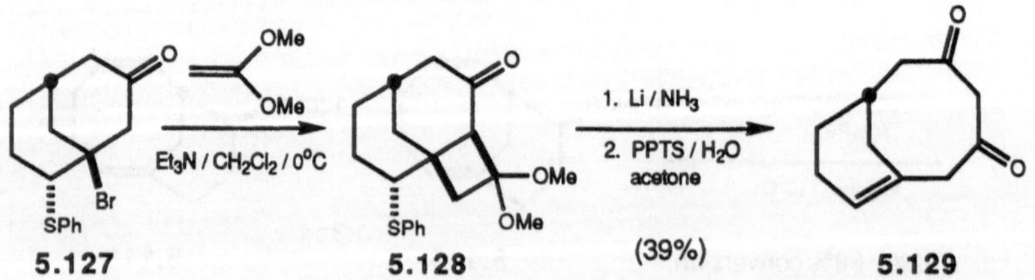

Scheme 5.24. Kraus' approach to bicyclo[5.3.1]undedecene ring systems

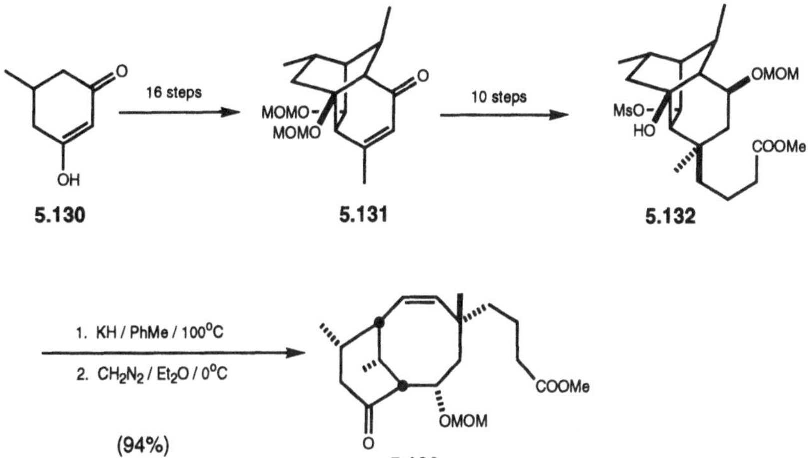

Scheme 5.25. Yamada's preparation of an AB intermediate

around a Grob fragmentation (Scheme 5.25). Starting with 5-methyl-1,3-cyclohexanedione (**5.130**) he prepared the key fragmentation substrate (**5.132**) via intermediate (**5.131**) in twenty-six steps. When (**5.132**) was treated with KH at 100°C, fragmentation occurred to give (**5.133**) in 94% yield, after reesterification. The bicyclo[5.3.1]undecane derivative (**5.133**) was intended to be an intermediate in the synthesis of taxinine, but, to date there have been no further reports.

5.1.3.9. Fétizon's Third Approach

The most recent report (*11*) by FÉTIZON's group describes a new approach that is strategically different from either of their first two (see 5.1.3.4 and 5.2.1.2). Like their second approach, this work focuses on the preparation of an AB intermediate that requires C-ring annulation for further conversion to taxoids. This approach began with the conversion of 2,6-dimethyl benzoquinone (**5.135**) and diene (**5.134**) into hydroxyester (**5.136**) (Scheme 5.26). Then using a two step procedure, hydroxyester (**5.136**) was transformed into key intermediate (**5.137**). Upon irradiation, tetracyclic hemiacetal (**5.137**) undergoes a Norrish II cleavage to give the unsaturated aldehyde (**5.138**) which is then converted to the taxoid precursor (**5.139**).

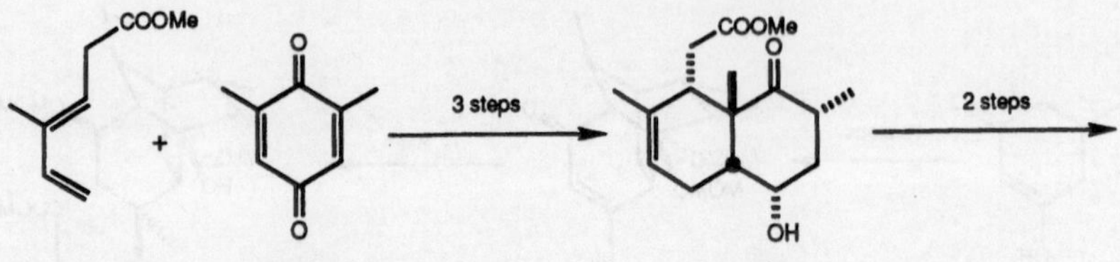

5.134   5.135   5.136

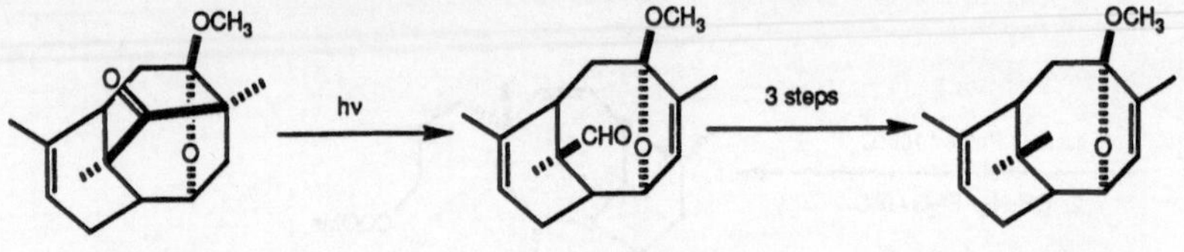

5.137   5.138   5.139

Scheme 5.26. Fétizon's third approach to the taxoids

5.1.3.10. Gadwood's Approach

Although it appears that GADWOOD has not pursued his 1982 strategy for the synthesis of taxoids, he was one of the earliest to investigate an approach involving an anionic oxy-Cope rearrangement (*80*). His efforts focused on the ring expansion of spiro[3.5]non-5-en-1-one (**5.140**) to various bicyclo[5.3.1]undec-1 (11)-en-4-ones (Scheme 5.27). Using a two step sequence, GADWOOD prepared the two AB taxoid models (**5.141**) and (**5.142**).

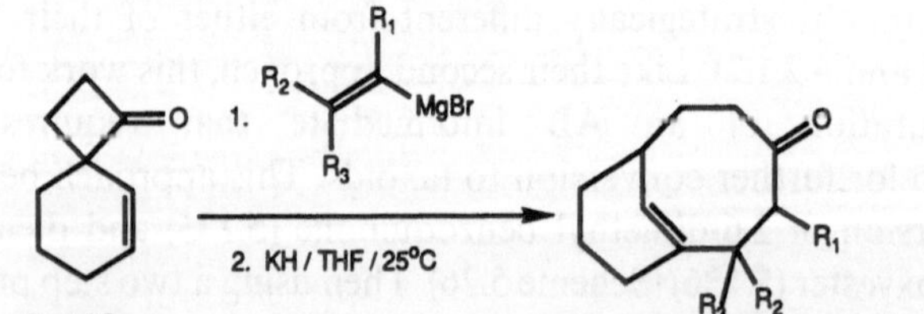

5.140

5.141 $R_1 = R_2 = R_3 = H$ (62%)
5.142 $R_1 = Me$ $R_2 = R_3 = H$ (49%)

Scheme 5.27. Gadwood's synthesis of bicyclo[5.3.1]undec-1(11)-en-4-ones

5.1.4. BC → ABC Approaches

5.1.4.1. Swindell's Approach

The approach of SWINDELL's group was based on the belief that a four carbon A-ring progenitor could be annulated onto a suitably function-alized BC ring intermediate (258). This belief was supported by MM2 calculations on a series of structures related to the taxoids which suggested that late attachment of the A-ring with direct introduction of the bridgehead double bond would be an energetically viable process (257). The construction of the BC-ring was to be accomplished using [2 + 2] photoaddition chemistry to join the B- and C-ring precursors followed by fragmentation chemistry to create a BC-ring intermediate with suitable functionality for A-ring annulation.

Initially, SWINDELL *et al.* explored the route shown in Scheme 5.28 for the preparation of BC intermediates (258, 262). Photoproduct (**5.143**) (231), prepared from dimedone and cyclohexanone cyanohydrin using a five step procedure (231, 259), was converted to enone (**5.144**). An

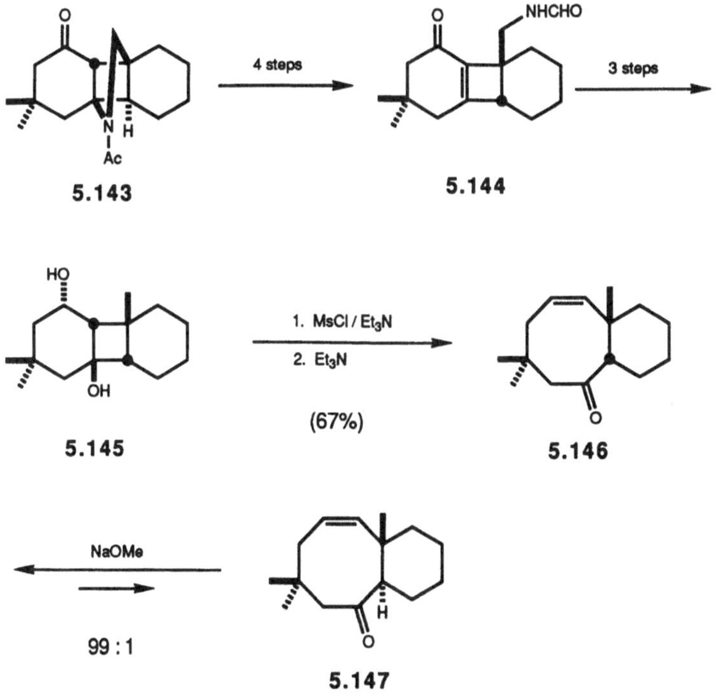

Scheme 5.28. Swindell's initial approach to the taxoid BC substructure

unexpected result of this conversion was epimerization at C-3 to give a *cis*-fused system at the ring fusion that would ultimately become the BC-ring fusion. In hopes that *trans*-fusion would be preferred at the bicyclo[6.4.0] stage the investigation was continued. Intermediate **(5.144)** was transformed into diol **(5.145)** which upon fragmentation yielded the BC intermediate **(5.146)**. An attempt to epimerize **(5.146)** at this point led to the discovery that it and not its C-3 epimer **(5.147)** had the thermodynamically preferred ring fusion stereochemistry. In light of this result, it was clear that the ring fusion stereochemistry in less stable *trans*-fused BC intermediate **(5.147)** would need to be derived directly from the corresponding ring fusion in the photoproduct **(5.143)**.

To this end, they examined the sequence outlined in Scheme 5.29 (*262*). A four step procedure provided access to the starting vinylogous imide **(5.148)** from dimedone and cyclohexanone cyanohydrin in 65% yield. Next, irradiation of **(5.148)** gave **(5.149)** which was in turn converted to fragmentation substrate **(5.150)**. Fragmentation of **(5.150)** led to intermediate **(5.151)** which was then transformed into the desired *trans*-fused BC intermediate **(5.147)**. From commercially available starting

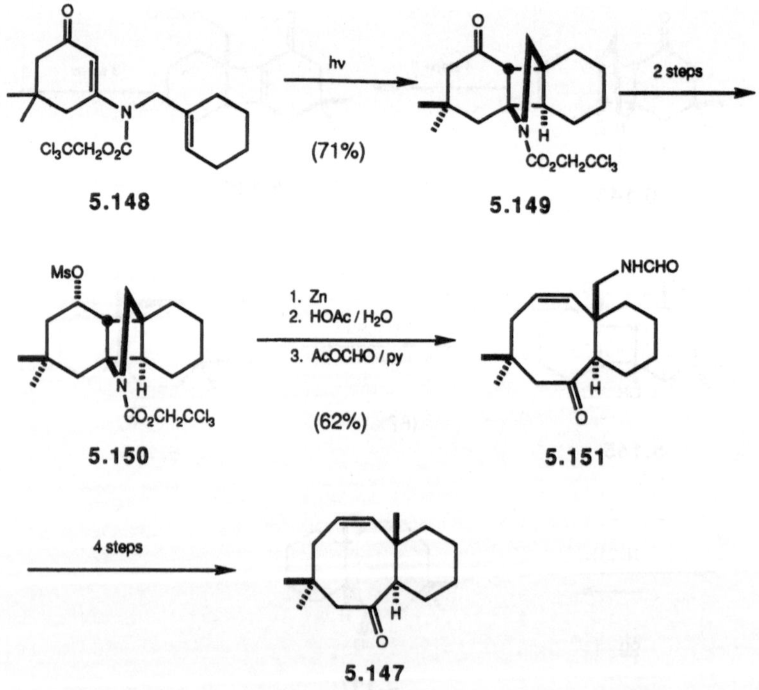

Scheme 5.29. Swindell's modified approach to the taxoid BC substructure

References, pp. 173–189

materials, the BC intermediate (**5.147**) was prepared in fourteen steps with a 21% overall yield.

With the crucial BC intermediate now in hand, SWINDELL and PATEL directed their efforts toward the development of methodology for annulation of the A-ring. Their initial target was the dihydrotaxoid (**5.154**) (*260*) (Scheme 5.30). Starting with photoproduct (**5.149**), key fragmentation substrate (**5.152**) was produced in six steps. Application of the fragmentation sequence converted (**5.152**) to bicyclic intermediate (**5.153**). A three step procedure was then used to cyclize the A-ring and form the tricyclic model (**5.154**). This twenty-two step route produced the dihydrotaxoid model (**5.154**) in 17% overall yield.

SWINDELL's most recent disclosure (*261*) described the application of this strategy to the construction of a tricyclic model which possesses an AB substructure equivalent to taxinine (Scheme 5.31). As with previous work, this begins with photoproduct (**5.149**). Formation of the siloxy ketone followed by SWINDELL's standard fragmentation methodology converted photoproduct (**5.149**) into (**5.155**). The next seven step sequence, designed to introduce the carbons that would ultimately become the A-ring, transformed (**5.155**) into key intermediate (**5.156**). Using this pivotal intermediate, three methods for completing the annulation of the A-ring were explored. This resulted in the preparation of the three tricyclic models (**5.157**)–(**5.159**). The most complex and significant was (**5.157**) whose AB substructure is identical to that of taxinine. The preparation of (**5.157**) was accomplished in 21 steps in 3.2% overall yield.

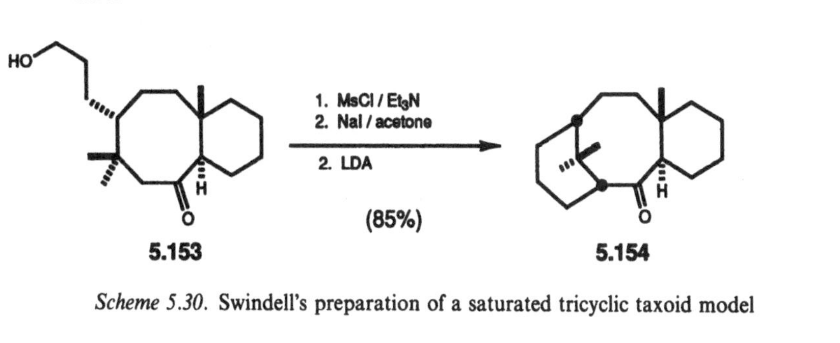

Scheme 5.30. Swindell's preparation of a saturated tricyclic taxoid model

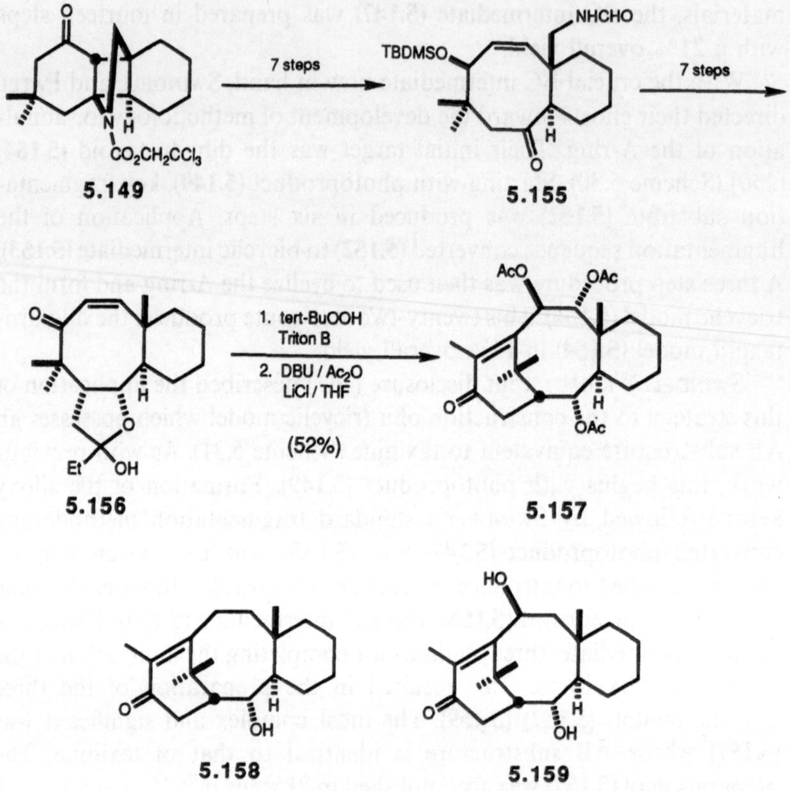

Scheme 5.31. Swindell's preparation of the taxinine AB system

5.1.4.2. Wender's A-Ring Annulation Approach

In addition to the formation of AB intermediates (see 5.1.3.6), WENDER has applied his nickel catalyzed intramolecular [4 + 4] cycloaddition methodology to the formation of BC intermediates. The initial report (287) disclosing this new methodology described the preparation of (5.163) from sorbic acid (5.161) and 1-bromohepta-4,6-diene (5.160) via tetraene (5.162) (Scheme 5.32). Similarly, the diastereomeric bicyclo [6.4.0] systems (5.167) and (5.168) were prepared from 4,6-heptadienal (5.164) and pentadienyl lithium (5.165) via tetraene (5.166). The impressive diastereoselectivity observed in the formation of (5.163) prompted WENDER to continue probing this new class of reactions in an effort to define the origins of stereoinduction and to develop a predictive model for general use (288).

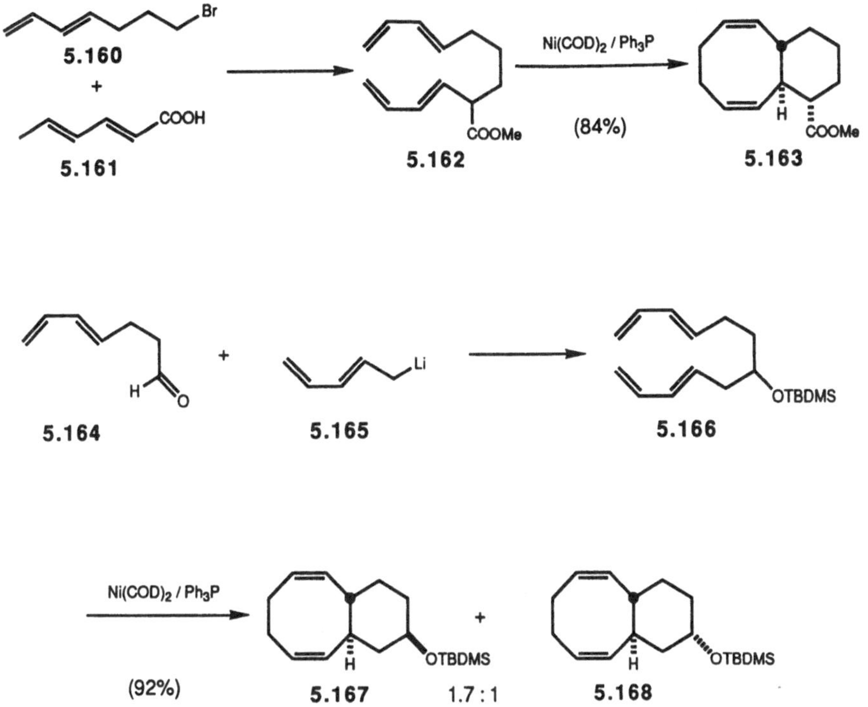

Scheme 5.32. Wender's initial synthesis of BC bicyclic systems

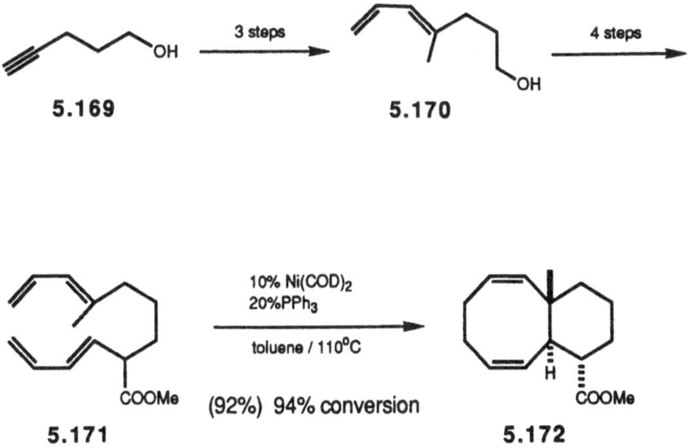

Scheme 5.33. Wender's BC bicyclization strategy for taxoid synthesis

With this background, WENDER directed his efforts toward the preparation of an angularly alkyl-substituted AB model (*290*) (Scheme 5.33). The cycloaddition precursor (**5.171**) was prepared from commercially available 4-pentyn-1-ol (**5.169**) in eight steps via intermediate dienol (**5.170**). Application of the cycloaddition methodology provided the BC intermediate (**5.172**) in 92% yield at 94% conversion with 97% diastereoselectivity.

5.1.4.3. Sieburth's Approach

SIEBURTH and CHEN recently disclosed a new approach to the preparation of 8-5 and 8-6 carbocyclic ring systems based on the intramolecular [4 + 4] photocycloaddition of tethered 2-pyridones (*247*). As a demonstration of the possible synthetic utility in the construction of taxoids, they prepared a *trans*-fused 8-6 ring system that may be viewed as a BC intermediate (Scheme 5.34). The key cycloaddition precursor (**5.175**) was prepared in 8 steps from 2,6-dibromopyridine (**5.173**) via intermediate (**5.174**). Upon irradiation, cycloadduct (**5.176**) was produced as an epimeric mixture (2:3) of carbinols in 63% yield.

5.1.5. Miscellaneous Approaches

Two linear approaches to the taxoid skeleton that have been reported do not conveniently fit into any of the previous catagories.

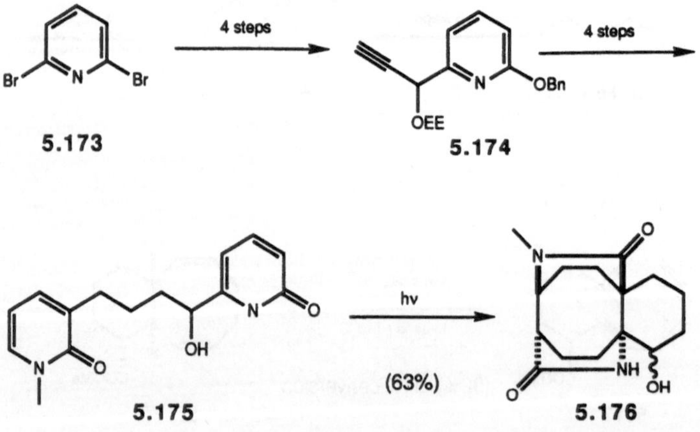

Scheme 5.34. Sieburth's preparation of a BC taxoid model

5.1.5.1. Kanematsu's Approach

KANEMATSU and coworkers have described an entry into bicyclo-
[n.3.1] ring systems that relies on the tandem intramolecular [2 + 2]
cycloaddition-[3,3] sigmatropic rearrangement of allenyl ethers (*100*).
The tricyclic[9.3.1.0^{4,9}]pentadecane systems (**5.181**) and (**5.182**) were
prepared (Scheme 5.35) as a demonstration of the synthetic utility this
new method might have in the preparation of taxoids. Starting from the
Wieland-Mischer ketones (**5.177**) and (**5.178**), key intermediates (**5.179**)
and (**5.180**) were prepared in thirteen steps. Exposure of either (**5.179**) or
(**5.180**) to base resulted in rearrangement to afford the tricyclic models
(**5.181**) and (**5.182**), respectively.

5.1.5.2. Kuwajima's Approach

As in the case of SHEA (see 5.1.2.1), the initial target of KUWAJIMA and
coworkers was a C-aromatic taxoid model (*111*). Their approach was
centered around the condensation reaction of acetals with cyclic enones
bearing (trimethylsilyl)methyl groups at their 3-position. Earlier work

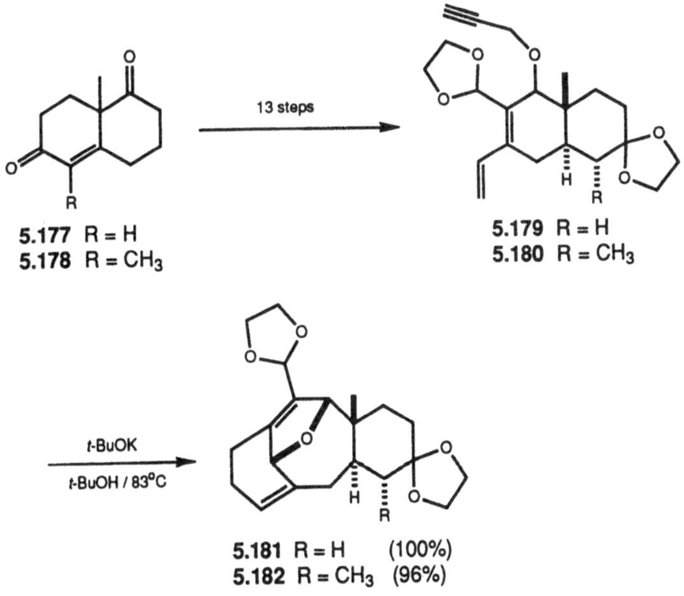

5.177 R = H
5.178 R = CH₃

5.179 R = H
5.180 R = CH₃

13 steps

t-BuOK
t-BuOH / 83°C

5.181 R = H   (100%)
5.182 R = CH₃   (96%)

Scheme 5.35. Kanematsu's preparation of tricyclic models

(99) had shown that condensation occurs exclusively at the methylene site bearing the trimethyl silyl group when SnCl₄ was used.

The construction of C-aromatic taxoid model (5.186) started with acetal (5.183) (Scheme 5.36). This was transformed into the key cyclization intermediate (5.185) in eight steps via enone (5.184). Upon treatment with TiCl₄-SnCl₄, (5.185) underwent cyclization to produce (5.186) in 73% yield.

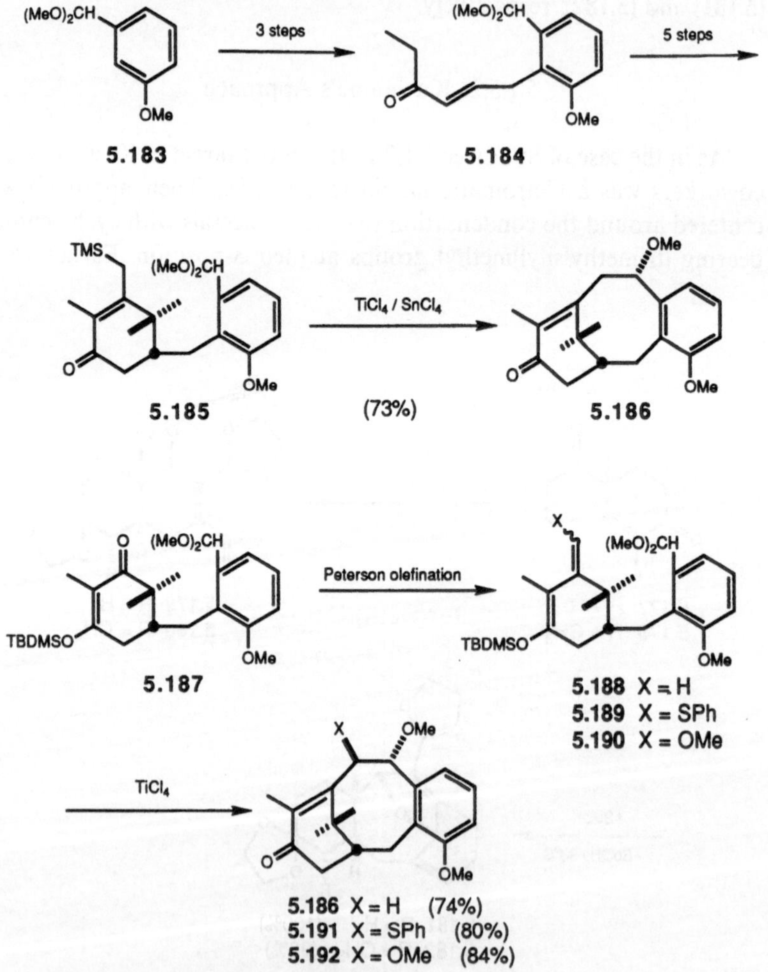

Scheme 5.36. Kuwajima's preparation of a C-aromatic tricyclic models

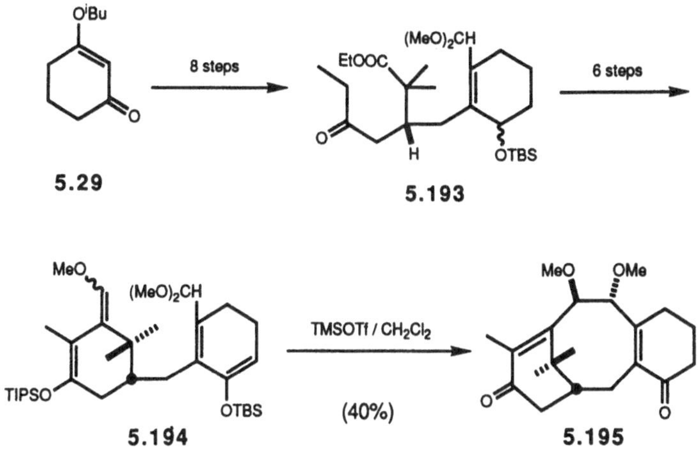

Scheme 5.37. Kuwajima's preparation of a partially saturated C-ring tricyclic model

Once this was achieved, the investigation was extended to include the introduction of functionality at C-10. Using a slight modification of the previously used methodology, enol silyl ethers (**5.188**)–(**5.190**) were prepared from (**5.187**), an intermediate from the initial route. On exposure to TiCl₄ cyclization occurred to give the corresponding eight membered ring products (**5.186**), (**5.191**), and (**5.192**) in good yields. Interestingly, (**5.189**) and (**5.190**), which were mixtures of E/Z isomers, produced a single stereoisomer. The origin of this stereochemical control was found to be thermodynamic in nature. The reaction of (**5.190**) at low temperature initially produced a mixture of two stereoisomers, but upon warming (**5.192**) was obtained as the sole product.

The most recent report from this group (*79*) describes the application of this methodology to a partially saturated C-ring tricyclic model (Scheme 5.37). In this case, the key intermediate (**5.194**) was synthesized from (**5.29**) via acetal (**5.193**). Exposure of (**5.194**) to TMSOTf resulted in cyclization to give the tricyclic taxoid model (**5.195**). As was previously observed, (**5.195**) was produced as a single stereoisomer.

5.2. Convergent Strategies

5.2.1. AC → ABC Approaches

This section contains approaches where the A- and C-ring precursors have been joined with the intention of forming the B-ring by direct closure. This type of approach has proven to be difficult.

5.2.1.1. Kitagawa's Approach

The efforts of KITAGAWA and coworkers in the area of taxoid synthesis have focused on the synthesis of optically active A- and C-ring intermediates. The preparation of **(5.201)** and **(5.202)**, two A-ring intermediates, was reported (*138, 142*) in 1984. The route that led to both intermediates is outlined in Scheme 5.38. This sequence began with the six step conversion of *d*-camphor **(5.196)** into the vinyl-cyclopentane derivative **(5.197)** (*139, 140*). Treatment of **(5.197)** with 2,4,4,6-tetrabromocyclohexa-2,5-dienone (TBCO) resulted in ring enlargement

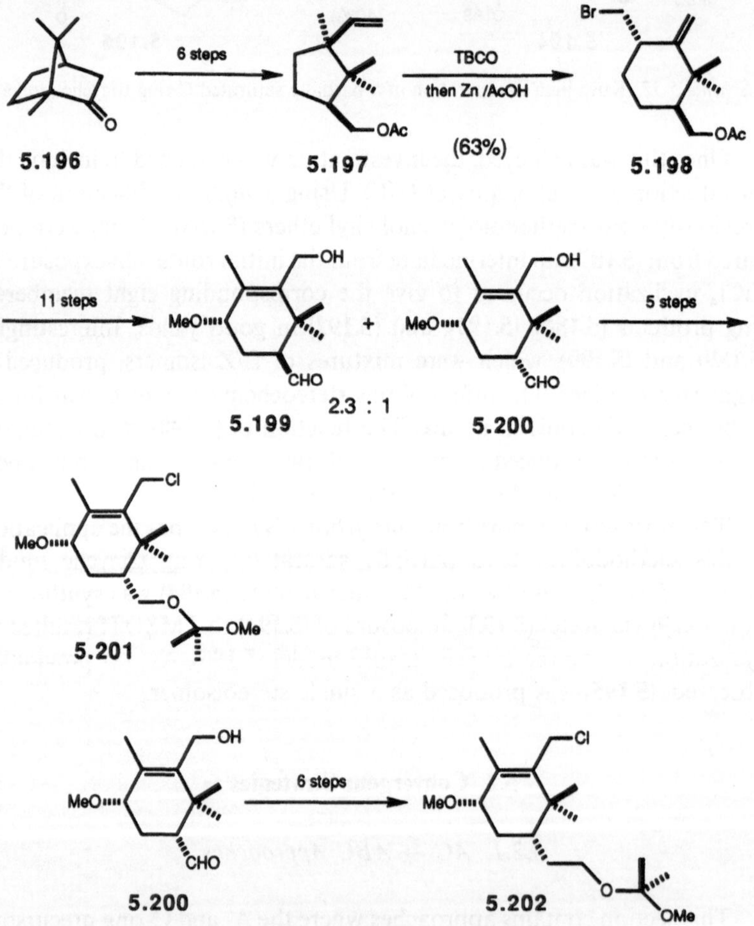

Scheme 5.38. Kitagawa's synthesis of two A-ring intermediates

to give bromomethyl-cyclohexane derivative (**5.198**) (*137*). The next series of steps led to an unfavorable mixture of aldehydes (**5.199**) and (**5.200**). After separation, the aldehyde with the desired configuration, (**5.200**), was converted to A-ring intermediate (**5.201**) in five steps. The second A-ring intermediate was then prepared by using a six step sequence to transform aldehyde (**5.200**) into (**5.202**).

Concurrently with his efforts to construct A-ring intermediates KITAGAWA *et al.* also explored the preparation of C-ring intermediates. Two approaches have been reported (*245*) for the synthesis of his initial C-ring intermediate, (**5.206**) (Scheme 5.39). The first route began with the conversion of 3-methylcyclohex-2-enone (**5.203**) into the diastereomeric mixture of (2*S*, 3*S*)-2,3-butanediol ketal (**5.204**). After separation, the desired diastereomer of (**5.204**) was transformed into the optically active C-ring intermediate (**5.206**) via intermediate ketal (**5.205**). The overall yield of (**5.206**) from 3-methylcyclohex-2-enone (**5.203**) was 15%.

Although the above route was successful, it did not lend itself to large scale application. Therefore, a second route was devised. This too, began with 3-methylcyclohex-2-enone (**5.203**). However, in this case (**5.203**) was converted to a diastereomeric mixture of (2*S*, 3*S*)-2,3-butanediol ketal (**5.208**) via intermediate ester (**5.207**). After separation, a four step

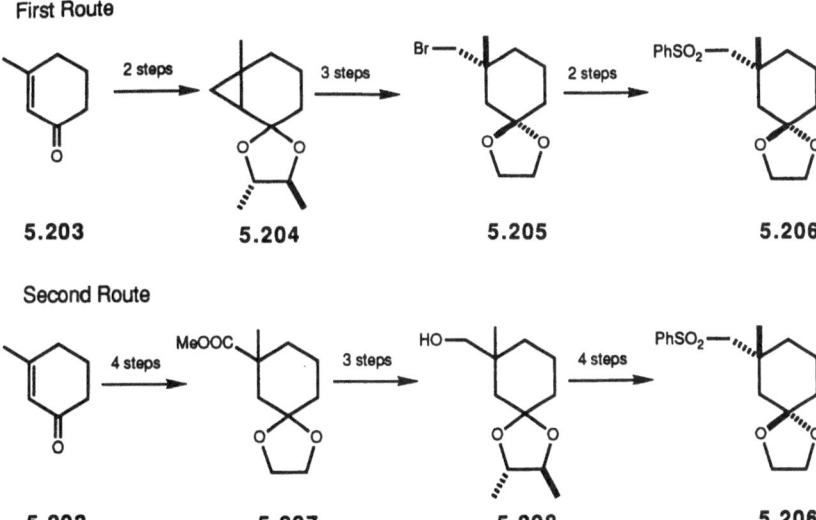

Scheme 5.39. Kitagawa's synthesis of C-ring intermediates

sequence again led to (5.206). This procedure produced the C-ring intermediate (5.206) in 12% overall yield.

Starting with the optically active keto-alcohol (5.209) which had been prepared in the latter route KITAGAWA and coworkers synthesized the alternate C-ring intermediate (5.212) (245) (Scheme 5.40). This route passed through the double bond isomers (5.210) and (5.211). In light of the fact that the undesired isomer (5.210) was the major product KITAG-AWA also developed a method to recycle it.

Although experimental details are not available, the KITAGAWA group has also reported the preparation of cyclization intermediates (5.214) (141) and (5.216) (246) (Scheme 5.41). Apparently, (5.214) was prepared from d-camphor (5.196) via intermediate (5.213). Cyclization intermediate (5.216) was constructed by coupling (5.201) and (5.206) followed by cyclization of the resulting intermediate, (5.215).

5.2.1.2. Fétizon's First Approach

The earliest report by FÉTIZON's group describe an attempt to prepare the taxoid skeleton by coupling A-and C-ring precursors followed by direct cyclization to form the B-ring (1). The investigation began with the preparation of C-ring precursor (5.217) (Scheme 5.42). This was prepared in four steps starting from 3-methylcyclohexenone (5.203). With this in hand, they directed their efforts toward the coupling of C-ring precursor (5.217) with diene (5.112) using a Mukaiyama condensation. Although a variety of conditions were tried, all attempts failed to produce the desired target, (5.218).

With the failure of the initial strategy, attention was directed to the examination of an alternate route. This approach involved the Michael condensation of the A- and C-ring precursors to form the lower portion of the B-ring. The initial test, the reaction of (5.112) with (5.217), resulted in the formation of the desired product, (5.219) (Scheme 5.43).

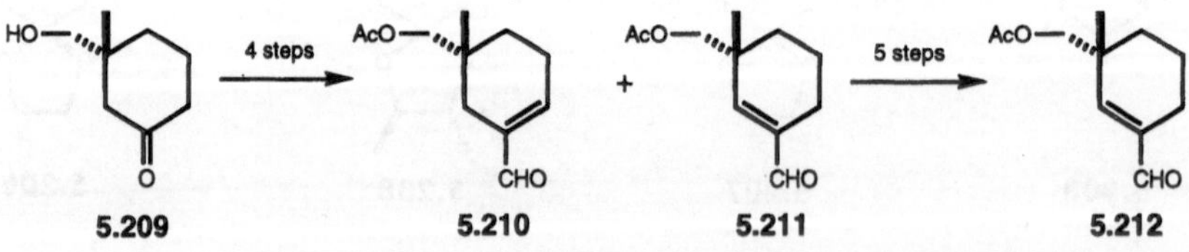

Scheme 5.40. Kitagawa's synthesis of an alternate C-ring intermediate

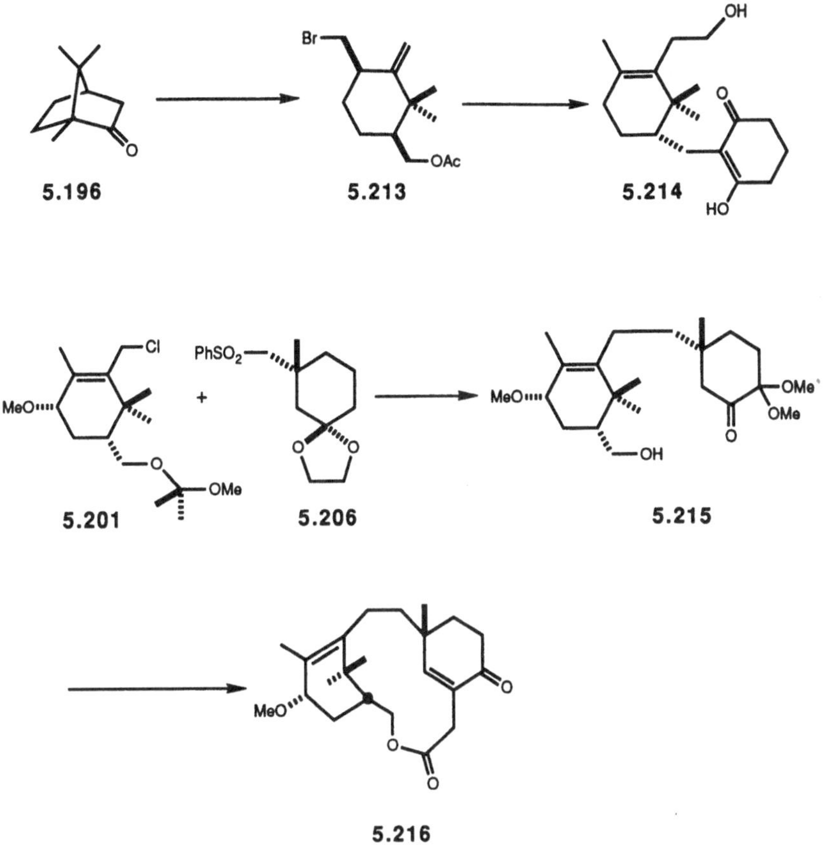

5.196   **5.213**   **5.214**

5.201   **5.206**   **5.215**

5.216

Scheme 5.41. Kitagawa's preparation of two taxoid skeleton precursors

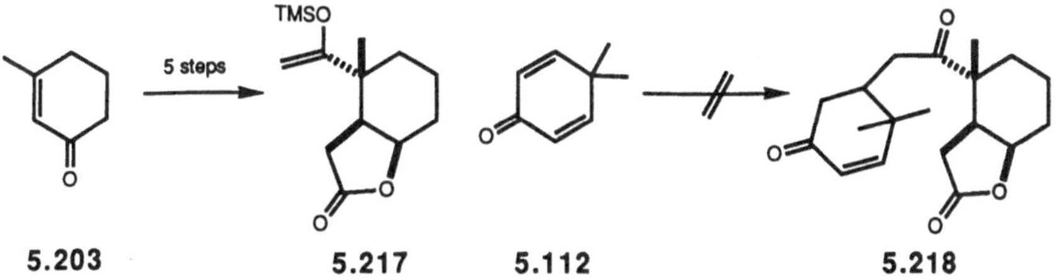

5.203   **5.217**   **5.112**   **5.218**

Scheme 5.42. Fétizon's first attempt at the taxoid skeleton

The success of the Michael reaction prompted the Fétizon group to explore the construction of the taxoid skeleton using the sequence outlined in Scheme 5.44. As with his first approach, this route begins with

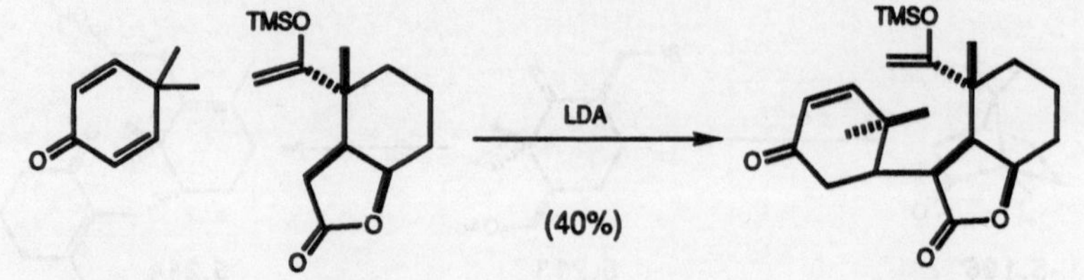

5.112 5.217 5.219

Scheme 5.43. A modification of Fétizon's first attempt at the taxoid skeleton

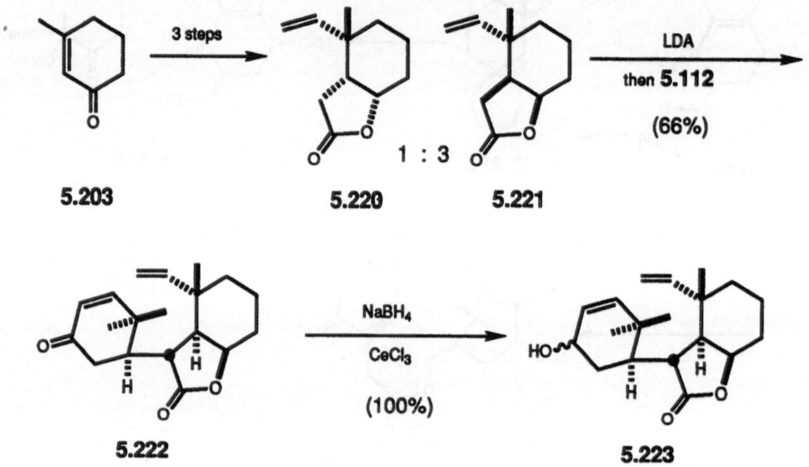

5.222 5.223

Scheme 5.44. Fétizon's alternate approach for direct cyclization

the conversion of 3-methylcyclohexenone (**5.203**) into the C-ring precursor. Using a three step sequence analogous to that previously used to synthesize the C-ring precursor, the inseparable mixture of diastereomers (**5.220**) and (**5.221**) were prepared. Michael condensation of this mixture with (**5.112**) led to key intermediate (**5.222**). Unfortunately, X-ray analysis revealed that the stereochemistry at C-1 was incorrect with respect to C-3/C-8. Furthermore, all attempts to induce cyclization of (**5.222**) or its reduction product (**5.223**) failed.

5.2.1.3. Kende's Approach

In 1986 KENDE *et al.* reported (*129*) the first total synthesis of a racemic tricyclic taxoid that possessed the full and stereochemically correct carbon framework of natural taxusin. Their strategy involved the coupling of A- and C-ring intermediates using a Mukaiyama aldol reaction followed by McMurry cyclization to form the sterically encumbered eight-membered B-ring.

The plan began with the construction of the A-ring intermediate (**5.225**) (Scheme 5.45) This was achieved in 21% overall yield using a ten step sequence that started with 2,6-dimethylcyclohexanone (**5.224**). Employing a Mukaiyama aldol process, acetal (**5.225**) was then coupled to C-ring intermediate (**5.226**) to yield enones (**5.227**) as a mixture of four isomers, two Z diastereoisomers and two E diastereoisomers. Next, a ten step sequence converted (**5.227**) into dialdehyde (**5.228**) which set the stage for closure of the B-ring. Exposure of (**5.228**) to McMurry Ti reagent from Zn-Cu and TiCl$_3$ produced cyclization products (**5.229**) and (**5.230**). Unfortunately, the desired (**5.230**) was the minor product. The last step converted (**5.230**) into taxoid triene (**5.231**). The twenty-four step route produced (**5.231**) in 0.1% overall yield.

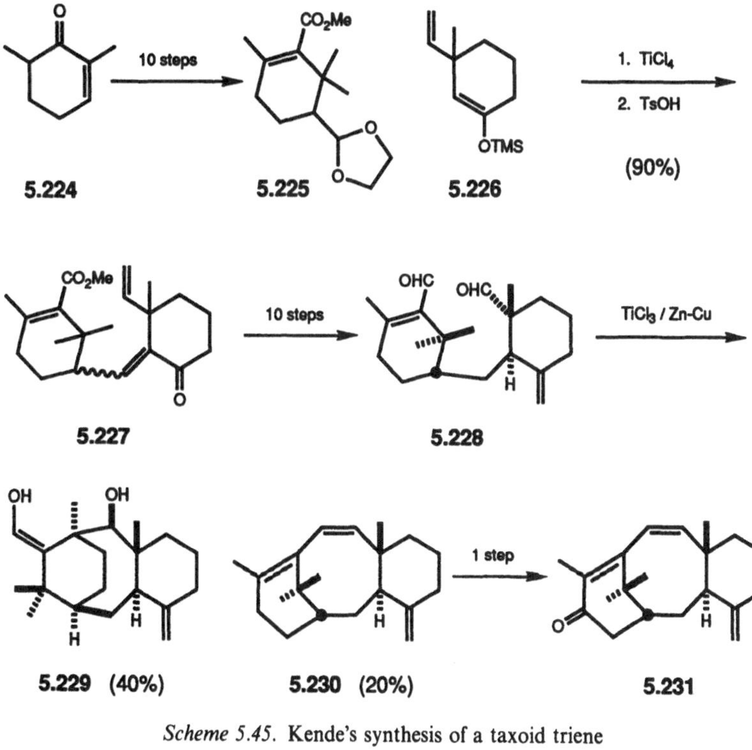

Scheme 5.45. Kende's synthesis of a taxoid triene

5.2.1.4. Funk's Approach

A strategy developed by FUNK and coworkers for the construction of the taxoid skeleton is based on methodology developed for the preparation of carbocycles using the Claisen rearrangement mediated contraction of macrocyclic lactones. Application of this methodology to the construction of the tricyclic taxoid ring system resulted in the tricyclic model (5.237) (78) (Scheme 5.46). Their approach began with the conversion of ketal alcohol (5.232) into the lithium dianion (5.233). This dianion was then coupled with aldehyde (5.235) which had been prepared from

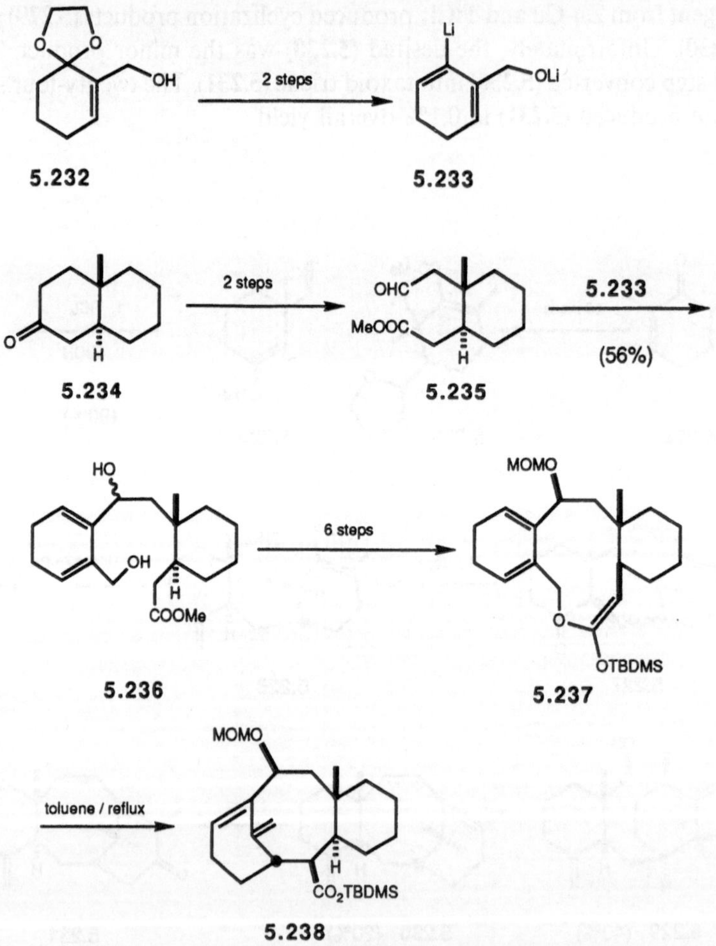

Scheme 5.46. Funk's synthesis of a tricyclic model

ketone (**5.234**) using a six step procedure. This coupling resulted in the formation of the diastereomeric mixture of diols (**5.236**) in 56% yield. The next series of steps which included the Mukaiyama method for formation of the macrocyclic lactone converted (**5.236**) into the rearrangement intermediate (**5.237**). Upon thermolysis, silyl ketene acetal (**5.237**) underwent rearrangement to yield the tricyclic taxoid model (**5.238**). This ten step construction of (**5.238**) demonstrated the utility of FUNK's Claisen rearrangement based methodology by providing the first efficient adaptation of a convergent strategy to the construction of a taxoid model.

5.2.2. A[B]C → ABC Approaches

In this type of approach, direct closure of the B-ring is avoided by joining A- and C-ring precursors in such a fashion that bond cleavage or reorganization of polycyclic intermediates will produce the eight-membered ring. The majority of convergent approaches fall in this category.

5.2.2.1. Trost's Approach

In 1982, TROST and HEIMSTRA disclosed (*276*) the details of an investigation that resulted in the preparation of AB intermediates (**5.244**) and (**5.247**) (Scheme 5.47). The critical step of his strategy was the fragmentation of intermediates such as (**5.242**), (**5.243**) and (**5.246**). These fragmentation imtermediates were prepared from 2-methylcyclohexenol (**5.239**). The route began with a six step sequence that produced a diastereomeric mixture of (**5.240**) and (**5.241**). After separation, (**5.240**) was converted to (**5.242**) while the same sequence transformed (**5.241**) into (**5.243**). Treatment of either (**5.242**) or (**5.243**) with a catalytic amount of *tert*-BuOK produced AB intermediate (**5.244**). The geminal dimethyl analogue (**5.246**) was prepared from (**5.245**) in two steps. This too, underwent fragmentation when exposed to a catalytic amount of *tert*-BuOK to yield (**5.247**).

Application of the same strategy led to the three step construction of AB intermediate (**5.254**) in 1984 (*278*) (Scheme 5.48). The synthetic route that led to (**5.254**) started with the condensation of allylic chloride (**5.248**) with enediol silyl ether (**5.249**). As was previously observed, this led to a 1:1 mixture of diastereomers (**5.250**) and (**5.251**). These were then transformed into their corresponding diols (**5.252**) and (**5.251**). Oxidative cleavage of either (**5.252**) of (**5.253**) resulted in the formation (**5.254**).

The next publication from this group (*277*) revealed their intention to employ this same strategy in the convergent synthesis of taxoids. If this

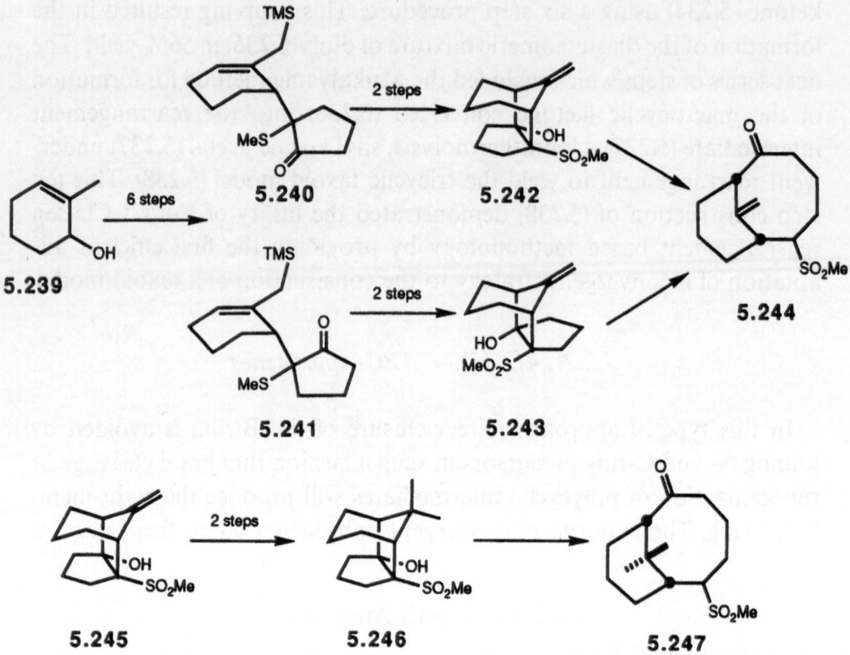

Scheme 5.47. Trost's initial approach for the construction of AB models

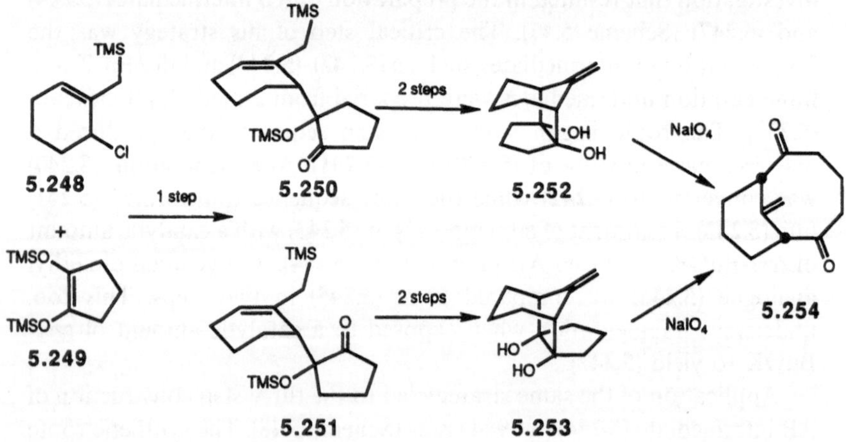

Scheme 5.48. Trost's modified route for the construction of AB models

were to be achieved, an appropriate *trans*-perhydroindanone, which would serve as a C-ring precursor, was required. Construction of this key intermediate relied on a ring contraction sequence via a modified benzilic acid rearrangement. They began by using a three step sequence to

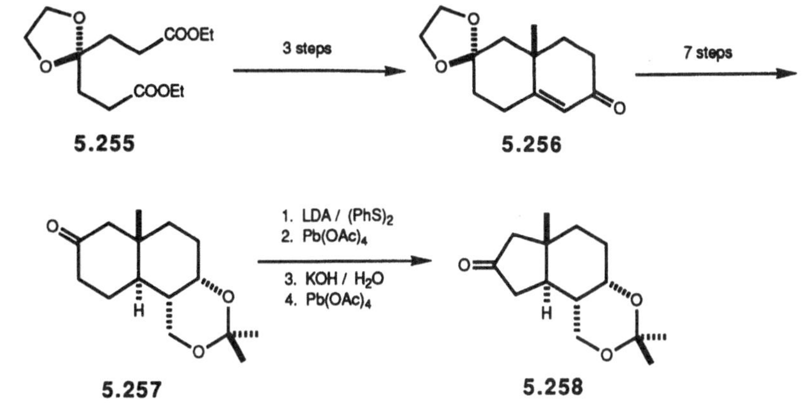

5.255   **5.256**

5.257   **5.258**

Scheme 5.49. Trost's preparation of a perhydroindanone

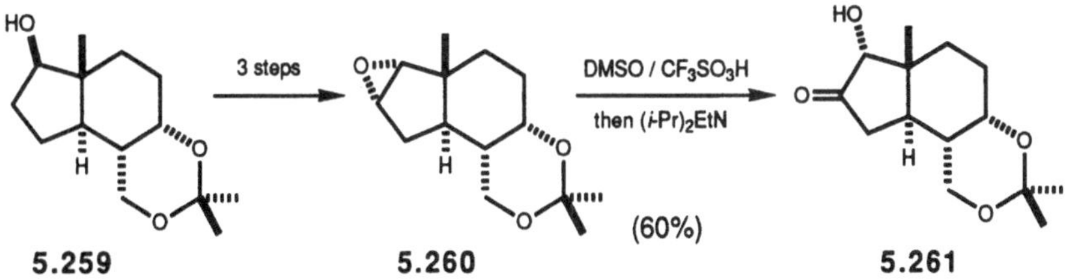

5.259   **5.260**   **5.261**

Scheme 5.50. Trost's preparation of an optically active perhydroindanone

convert the ethylene ketal of 4-oxopimelate (**5.255**) into octalone (**5.256**) (Scheme 5.49). The next series of steps transformed (**5.256**) into (**5.257**) which set the stage for his key ring contraction sequence. Application of this final sequence led to the desired perhydroindanone (**5.258**).

The most recent report (*275*) from this group describes the preparation of the optically active and more complex perhydroindanone intermediate (**5.261**) (Scheme 5.50). Application of his methodology based on the oxidative ring-opening of epoxides converted enantiomerically pure alcohol (**5.259**) into hydroxy ketone (**5.261**) via epoxide (**5.260**).

5.2.2.2. Inouye's First Approach

The first approach by INOUYE and coworkers was based on the bond disconnections shown in the retrosynthetic analysis of Scheme 5.51 (*116*). With this in mind, they explored a number of routes designed to prepare the 5-alkyl-1,5,11,11-tetramethyltricyclo[6.2.1.0^{2,6}]undecane system (**5.266**). After several unsuccessful attempts to prepare (**5.266**), they

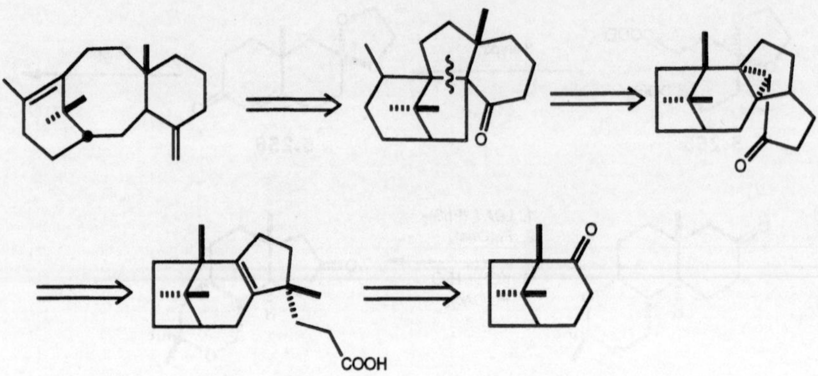

Scheme 5.51. Inouye's retrosynthetic analysis

succeeded in the synthesis of the two related 1,5,5,11,11-pentamethyltricyclo[6.2.1.0^{2,6}]undecane derivatives (**5.265**) and (**5.270**)

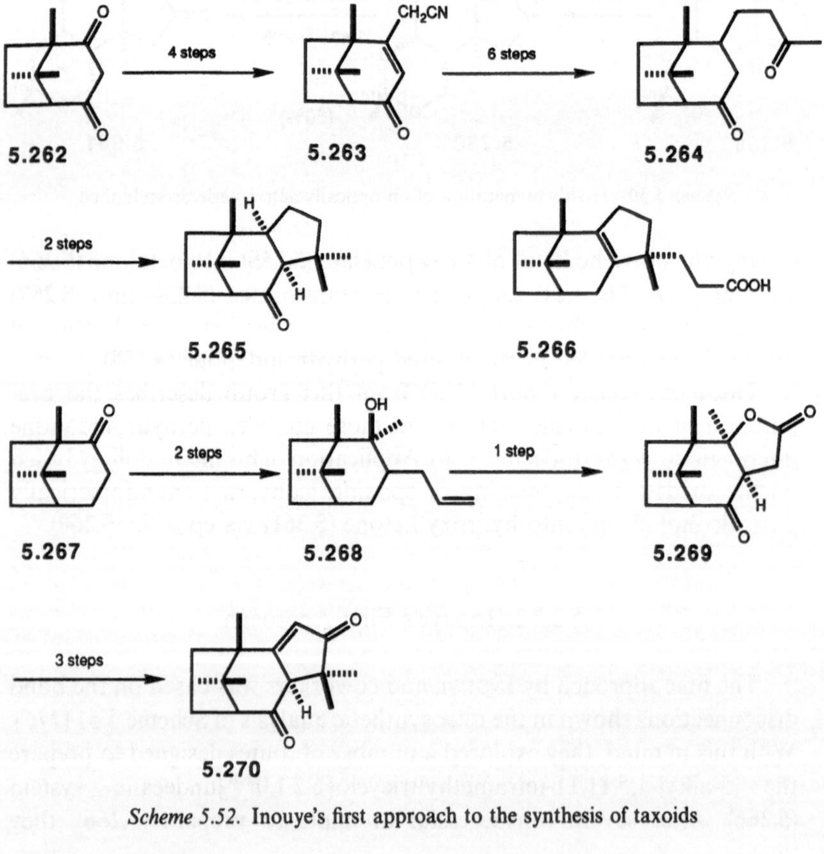

Scheme 5.52. Inouye's first approach to the synthesis of taxoids

(Scheme 5.52). The sequence that led to (**5.265**) began with diketone (**5.262**). This was converted to enone (**5.263**) in four steps which was then transformed into (**5.264**). Cyclization of (**5.264**) followed by conjugate addition produced (**5.265**). Construction of (**5.270**) involved the conversion of homocamphor (**5.267**) into alcohol (**5.268**). After conversion to lactone (**5.269**), a three step sequence produced (**5.270**). Apparently, INOUYE has abandoned this approach since recent reports describe a different strategy.

5.2.2.3. Blechert's First Approach

The first approach of BLECHERT's group like their second (see 5.1.3.5), relied on the de Mayo sequence for construction of the central eight-membered ring. However, unlike their second approach which created AB intermediates, the initial efforts were directed towards the construction of tricyclic models.

A 1984 report (*189*) described a stereoselective route for the preparation of the tricyclic model (**5.275**) (Scheme 5.53). Construction of this

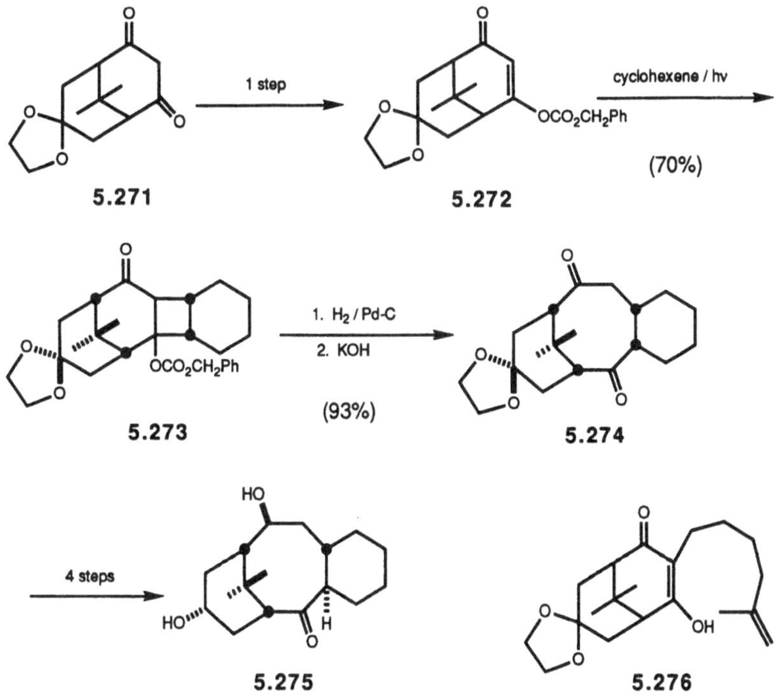

Scheme 5.53. Blechert's synthesis of a tricyclic taxoid model

model began with bicyclic diketone (**5.271**). After conversion to its benzyl carbonate derivative (**5.272**), intermolecular photocycloaddition resulted in (**5.273**). Deprotection followed by fragmentation led to tricycle (**5.274**). Although this intermediate possesses the correct carbon connectivity for the taxoid skeleton, the stereochemistry at C-3 is not correct. Fortunately, the final four step sequence allowed epimerization at C-3 to produce the tricyclic taxoid model (**5.275**).

Although this route was successful in the construction of (**5.275**), attempts to modify it to include a methyl group at C-8 proved less productive. The use of 1-methylcyclohexene in place of cyclohexene succeeded in generating the cyclobutane intermediate, but failed in terms of regioselectivity. Another plan to introduce a methyl group at the C-8 position involved the photocycloaddition of (**5.276**). Unfortunately, this failed to cyclize.

5.2.2.4. Clark's Approach

In 1987, CLARK and coworkers reported (*160*) the synthesis of a racemic A-ring fragment and a racemic C/D ring fragment that were to be coupled by an aldol reaction. In order for this strategy to succeed, the aldol reaction would have to proceed with mutual kinetic resolution. An earlier investigation by the same group (*41*) which focused on the aldol reaction of cyclobutanones indicated this approach might be feasible. This prompted them to pursue synthetic routes to A-ring precursor (**5.281**) and C/D ring precursor (**5.284**).

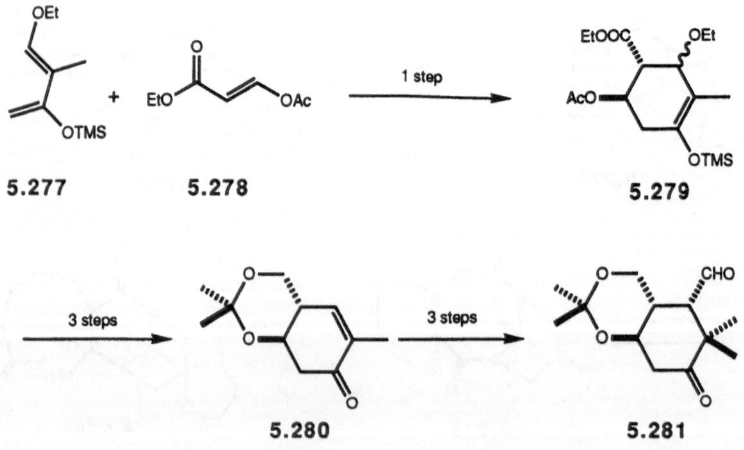

Scheme 5.54. Clark's synthesis of an A-ring precursor

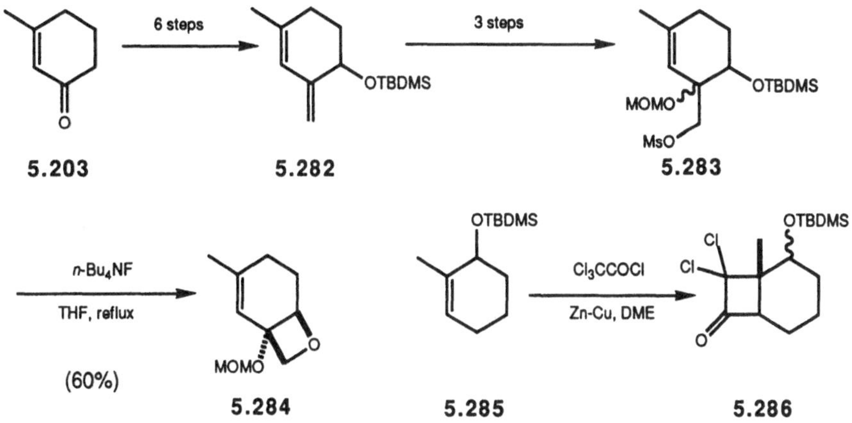

Scheme 5.55. Clark's synthesis of a C/D ring precursor

Their synthesis of A-ring precursor is outlined in Scheme 5.54. Diels-Alder reaction of (**5.277**) with (**5.278**) produced the highly functionalized cyclohexene (**5.279**) which was then elaborated to enone (**5.280**). The final three step sequence culminated in (**5.281**).

Preparation of the C/D ring precursor began with the six step conversion of 3-methyl-2-cyclohexen-1-one (**5.203**) into (**5.282**) (Scheme 5.55) The next series of steps converted (**5.282**) into mesylate (**5.283**) which on desilylation led to oxetane (**5.284**). They also performed model studies to investigate the formation of the requisite cyclobutanone. Cycloaddition of (**5.285**) with dichloroketene, prepared *in situ*, produced (**5.286**) in 30% yield.

5.2.2.5. Berkowitz's Approach

In 1985, BERKOWITZ and coworkers reported (*13, 14*) the results of an extensive investigation that employed the de Mayo sequence to produce the central eight-membered B-ring of the taxoid skeleton. Both inter-molecular and intramolecular photocycloadditions of cycloalkenes with homocamphorquinone derivatives were examined as well as a variety of conditions to fragment the photoadducts. In order to focus their attention on the development of photocycloaddition-fragmentation methodology, they chose to use the readily available homocamphorquin-one as a model rather than pursue the synthesis of an appropriate bicyclo[3.3.1]-nonane-2,4-dione system.

A series of intermolecular photocycloadditions using both cyclohex-ene and cyclopentene was examined initially in order to determine the stereochemical constraints. Some typical examples are shown in Scheme 5.56. Photoaddition of cyclopentene to camphorquinone derivative (**5.287**) resulted in the single cycloadduct (**5.288**). Upon fragmentation, (**5.289**) was produced in 60% yield along with the competing elimination

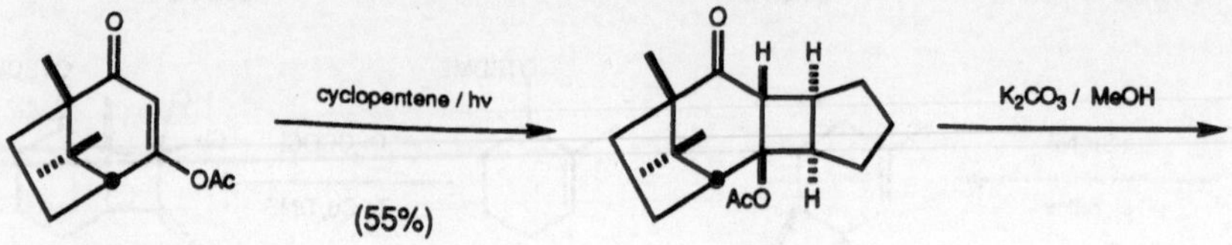

5.287 **5.288**

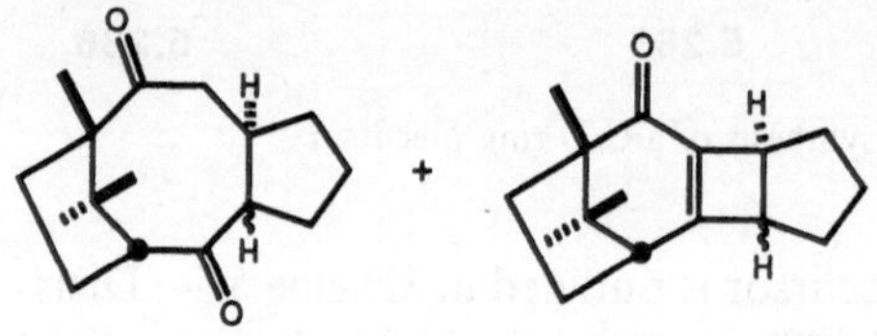

5.289 (60%) **5.290** (20%)

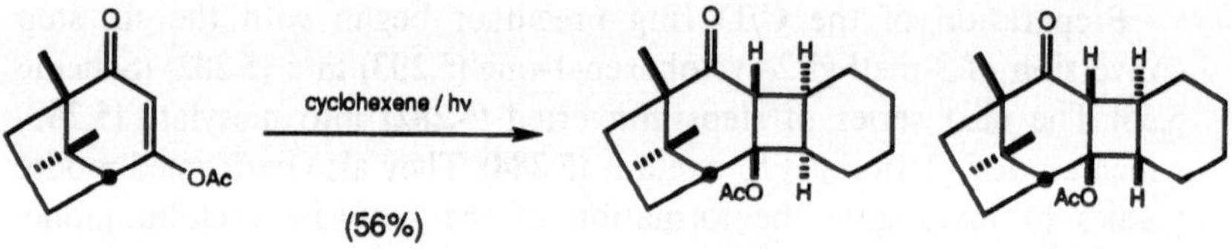

5.287 **5.291** **5.292**

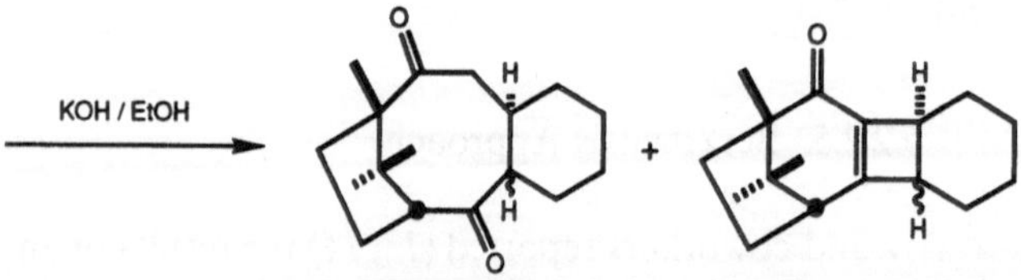

5.293 (60%) **5.294** (20%)

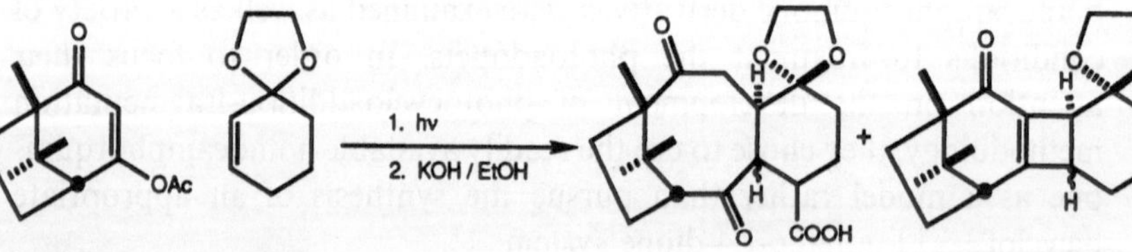

5.287 **5.295** **5.296** (40%) **5.297** (15%)

Scheme 5.56. Berkowitz's intermolecular photocycloaddition model studies product (**5.290**). In contrast, the reaction of homocamphor derivative (**5.287**) with cyclohexene led to a 1:1 mixture of (**5.291**) and (**5.292**). Fragmentation of the mixture under basic conditions led to (**5.293**) and (**5.294**) in 60% and 20% yield, respectively. Fragmentation of the same mixture under acidic conditions produced the same two products, but in lower yield. As the third example shows, they also investigated the photocycloaddition of a functionalized cyclohexene. When enol acetate (**5.287**) was subjected to photoaddition-fragmentation conditions with (**5.295**), the result was a mixture of tricycle (**5.296**) and elimination product (**5.297**).

The model studies of BERKOWITZ *et al.* also included two intramolecular photoaddition sequences. Initially, photocyclization of (**5.301**) was examined (Scheme 5.57). Photoaddition substrate (**5.301**) was prepared by coupling dimedone (**5.300**) and bromoketal (**5.299**). Upon irradiation (**5.301**) was converted to (**5.302**).

A second sequence involved the coupling of diketone (**5.262**) with bromoketal (**5.299**) (Scheme 5.58). This led to (**5.303**) which upon irradiation produced photoproduct (**5.304**). Subjection of the latter to fragmentation conditions gave the tricyclic model (**5.305**).

5.2.2.6. Inouye's Second Approach

The INOUYE group, like several others, has also investigated an approach that relies on intramolecular [2 + 2] photocycloaddition-

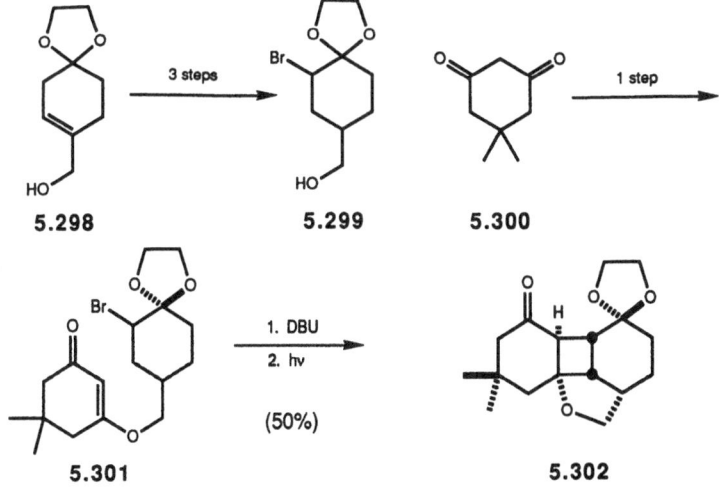

Scheme 5.57. Berkowitz's intramolecular photoaddition model study

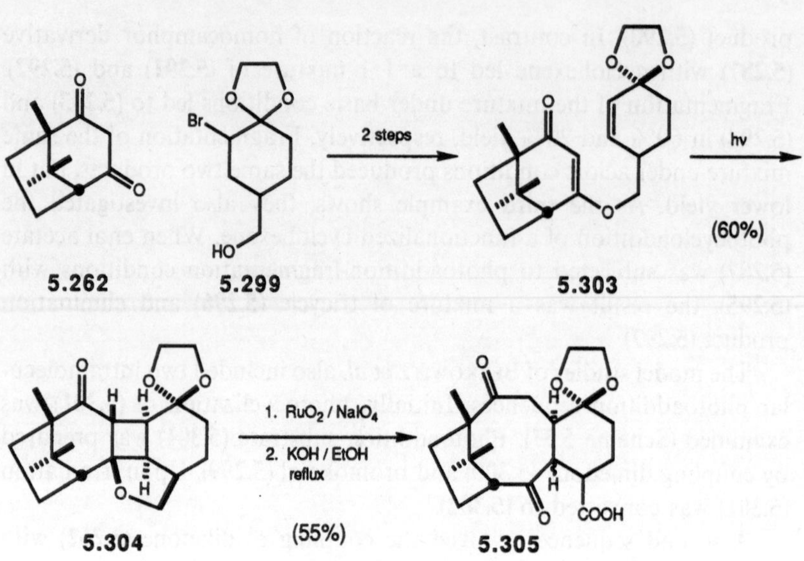

Scheme 5.58. Berkowitz's synthesis of a tricyclic model

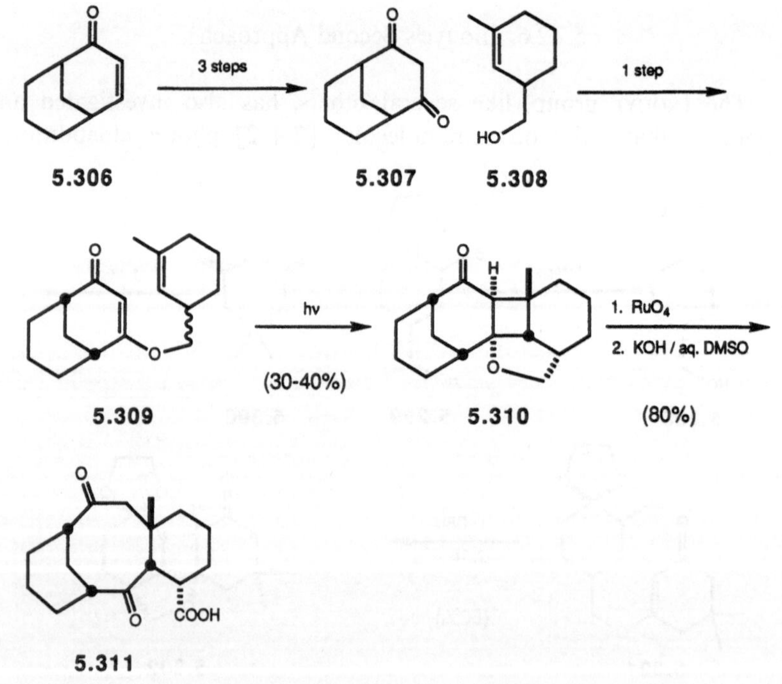

Scheme 5.59. Inouye's synthesis of a tricyclic taxoid model

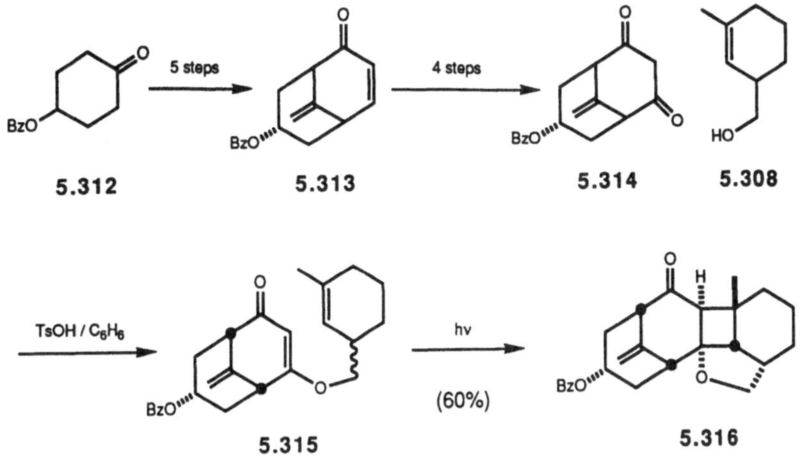

5.312   5.313   5.314   5.308

5.315   5.316

Scheme 5.60. Inouye's approach to a functionalized tricyclic model

fragmentation chemistry for the construction of the eight-membered ring. Their application of this methodology led to the tricyclic taxoid model (**5.311**) (*144*) (Scheme 5.59).

The construction of the tricyclic model (**5.311**) began with the three step conversion of enone (**5.306**) into diketone (**5.307**). Diketone (**5.307**) was then coupled to alcohol (**5.308**) to generate the diastereomeric mixture of enol ethers (**5.309**). Irradiation of (**5.309**) led to photoproduct (**5.310**) in 30–40% yield. The modest yield was due to the fact that only one diasteromer underwent photoaddition. Finally, a two step sequence was used to induce fragmentation which produced the tricyclic model (**5.311**).

The most recent publications (*117, 145*) describe the application of this same strategy to a more functionalized tricyclic model (Scheme 5.60). This endeavor began with the construction of the functionalized bicyclo [3.3.1] nonane derivative (**5.314**) from (**5.312**) using a nine step sequence that passed through intermediate (**5.313**) (*117*). As was done in the previous synthesis, diketone (**5.314**) was then coupled to alcohol (**5.308**) to give (**5.315**). Irradiation led to photoadduct (**5.316**) in 60% yield. Apparently this is as far as the work has progressed (*115*), so the results of the fragmentation of this photoadduct are unavailable.

5.2.2.7. Winkler's Approach

The application of intramolecular [2 + 2] photoaddition-fragmentation chemistry to the synthesis of taxoids by WINKLER and coworkers is distinctly different from other approaches of this type. Their strategy focused on the intramolecular photoaddition of dioxolenones to alkenes.

The initial report (*294, 297*) from this group described the application of his strategy to the synthesis of a bicyclo[5.3.1] ring system. Photosubstrate (**5.318**) was prepared from *tert*-butyl cyclohexanone-2-carboxylate (**5.317**) in two steps (Scheme 5.61). Irradiation resulted in photoproduct (**5.319**) which was directly subjected to fragmentation conditions to yield (**5.320**). The final three step sequence led to the *trans*-bicyclo[5.3.1]undecanone (**5.321**).

The next report (*295*) disclosed the details of an investigation that focused on the construction of a tricyclic taxoid model using the same strategy (Scheme 5.62). The substrate required for the intramolecular dioxolenone cycloaddition was prepared from keto ester (**5.322**). The result of this two step sequence was a 1:1 mixture of alkylated dioxolenones (**5.323**) and (**5.324**). After separation, (**5.324**) was irradiated to yield photoadduct (**5.325**). Unfortunately, exposure of (**5.325**) to his usual fragmentation conditions led to lactone (**5.326**) and not the expected tricyclic ketoester (**5.327**).

In 1989, Winkler *et al.* reported (*296*) the details of a successful application of the intramolecular dioxolenone photocycloaddition chemistry to the construction of a tricyclic taxoid model (Scheme 5.63). The requisite photosubstrate (**5.329**) was prepared in seven steps from ketal alcohol (**5.328**). Irradiation resulted in the photoadduct (**5.330**) in 77% yield. Subjection of (**5.330**) to fragmentation conditions resulted in a 3:1

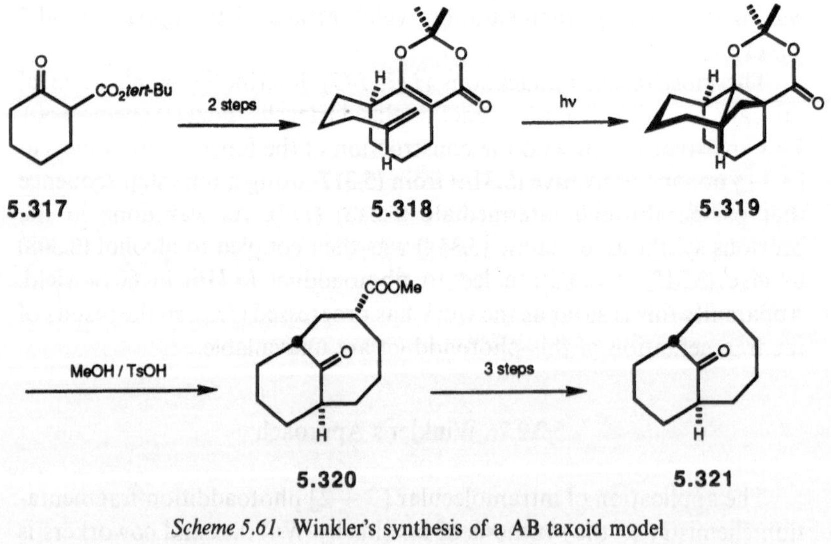

Scheme 5.61. Winkler's synthesis of a AB taxoid model

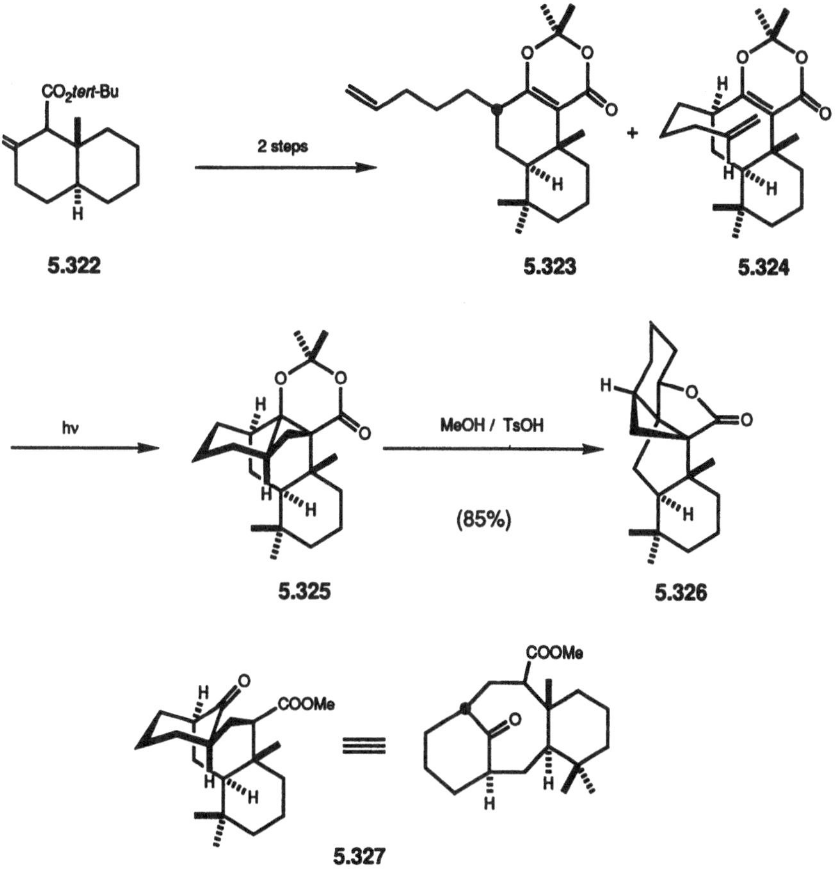

5.322 5.323 5.324 5.325 5.326 5.327

Scheme 5.62. Winkler's first attempt at a tricyclic taxoid model

mixture of (**5.331**) and its C-15 epimer. The final sequence which was applied to the major fragmentation product (**5.331**) led to the tricyclic taxoid model (**5.332**).

5.2.2.8. A Variation on Fétizon's Second Approach

During the course of their investigation that examined the use of the de Mayo sequence to construct the AB portion of the taxoid ring system (see 5.1.3.4), Fétizon's group also explored the application of this methodology to a convergent synthesis of the taxoid skeleton (*31*). The results of this brief investigation are shown in Scheme 5.64.

The photocycloaddition of enol acetate (**5.333**) was examined first. When (**5.333**) was irradiated in the presence of excess cyclohexene, the

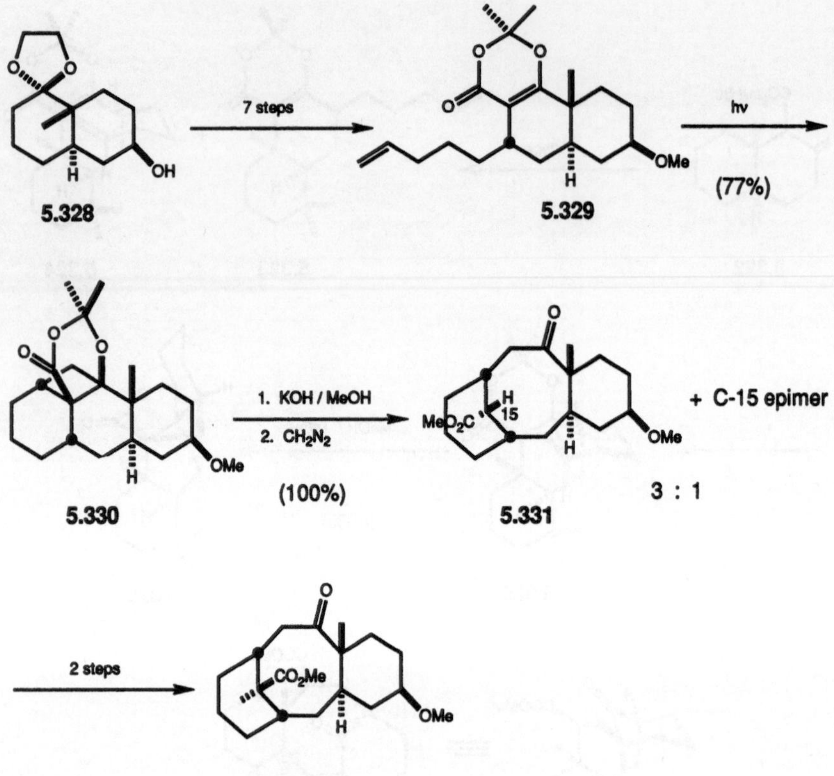

Scheme 5.63. Winkler's synthesis of a tricyclic taxoid model

result was photoproduct (**5.334**). Base catalyzed fragmentation of (**5.334**) led to a 3:2 mixture of the tricyclic model (**5.335**) and the unsaturated ketone (**5.336**). Although rings B and C of (**5.335**) are *trans*-fused, their relative stereochemistry with respect to the bridgehead protons is opposite that of the taxoid system.

Similar results were obtained when enol ether (**5.337**) was used. Irradiation led to a 3:1 mixture of photoproducts (**5.338**) and (**5.339**). Acid induced fragmentation of this mixture resulted in (**5.340**) in 55% yield plus unreacted (**5.338**).

5.2.2.9. Ghosh's Approach

In a 1991 article (*225*) Ghosh and coworkers outlined a strategy for the synthesis of the taxoid skeleton based on the fragmentation of

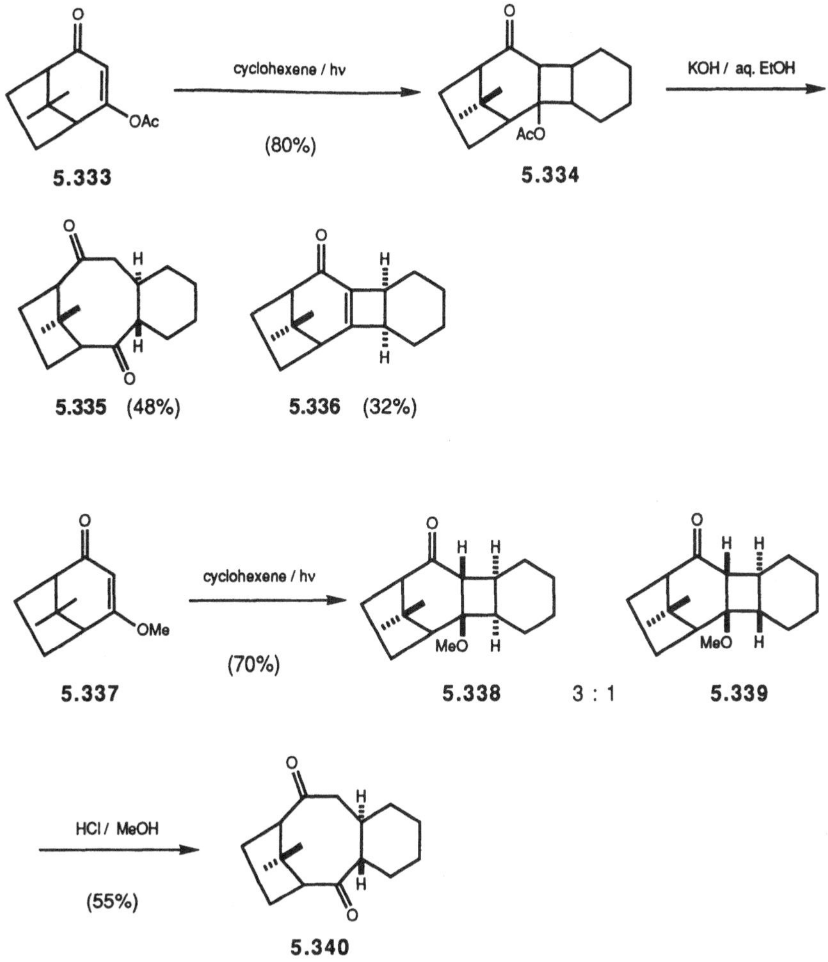

5.333   **5.334**

5.335 (48%)   **5.336** (32%)

5.337   **5.338**   3 : 1   **5.339**

5.340

Scheme 5.64. Fétizon's intermolecular photocycloaddition model studies

bicyclo[2.2.1]heptane derivatives. In order to demonstrate the feasibility of this approach, they prepared the bicyclo[5.3.1]undecane model (**5.344**) and the tricyclic diester (**5.347**) (Scheme 5.65). Construction of (**5.344**) began with a three step sequence to prepare the key fragmentation intermediate (**5.342**). Upon subjection to fragmentation conditions, the bicyclic diester (**5.343**) was produced. Transformation of the bicyclo [5.2.1]decane (**5.343**) into the taxoid AB model (**5.344**) was accomplished using a three step protocol. Application of this same methodology converted (**5.345**) to tricyclic model (**5.347**) via fragmentation intermediate (**5.346**).

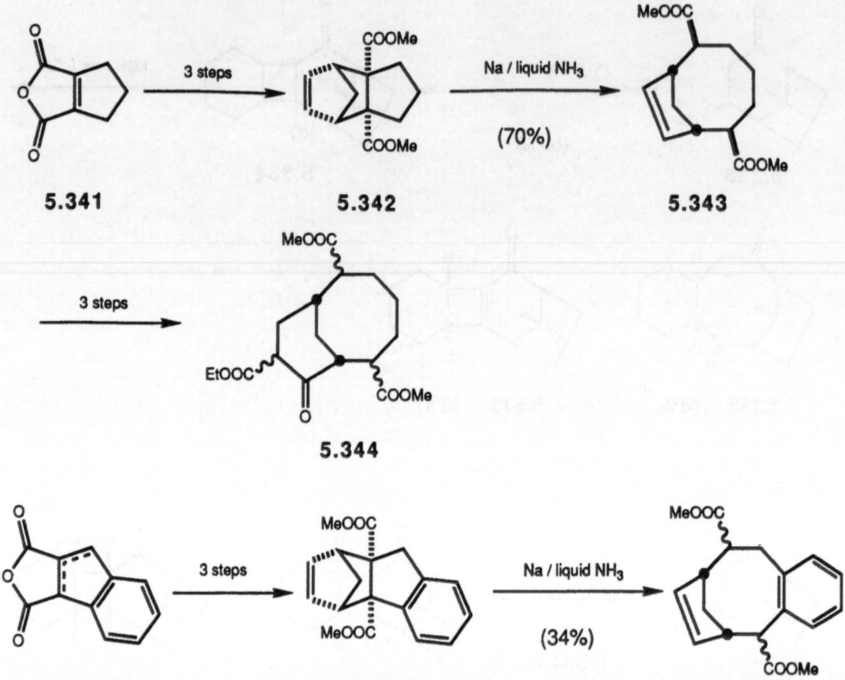

5.341   5.342   5.343

5.344

5.345   5.346   5.347

Scheme 5.65. Ghosh's fragmentation-based strategy for the taxoid skeleton

Most recently, this group has reported an improved method for the preparation of his key bicyclo[2.2.1]heptane intermediates (226).

5.2.2.10. Paquette's Approach

Two recent reports by PAQUETTE and coworkers (208, 209) describe the anionic oxy-Cope rearrangement of 1-vinyl-2-alkenyl-7,7-dimethyl-exo-norbornan-2-ols to create tricyclic systems. Although these tricycles do not possess carbon skeletons identical to that of the taxoids, if one envisioned the expansion of the A-ring in concert with contraction of the B-ring, it would be possible, in principle, to create the taxoid ring system. While these two reports did not clearly reveal an intent to use this methodology in the construction of taxoids, the most recent publication from this group (207) does substantiate this notion.

From this initial work, two sequences possess some relevance to the synthesis of taxoids. The first is outlined in Scheme 5.66. This route began

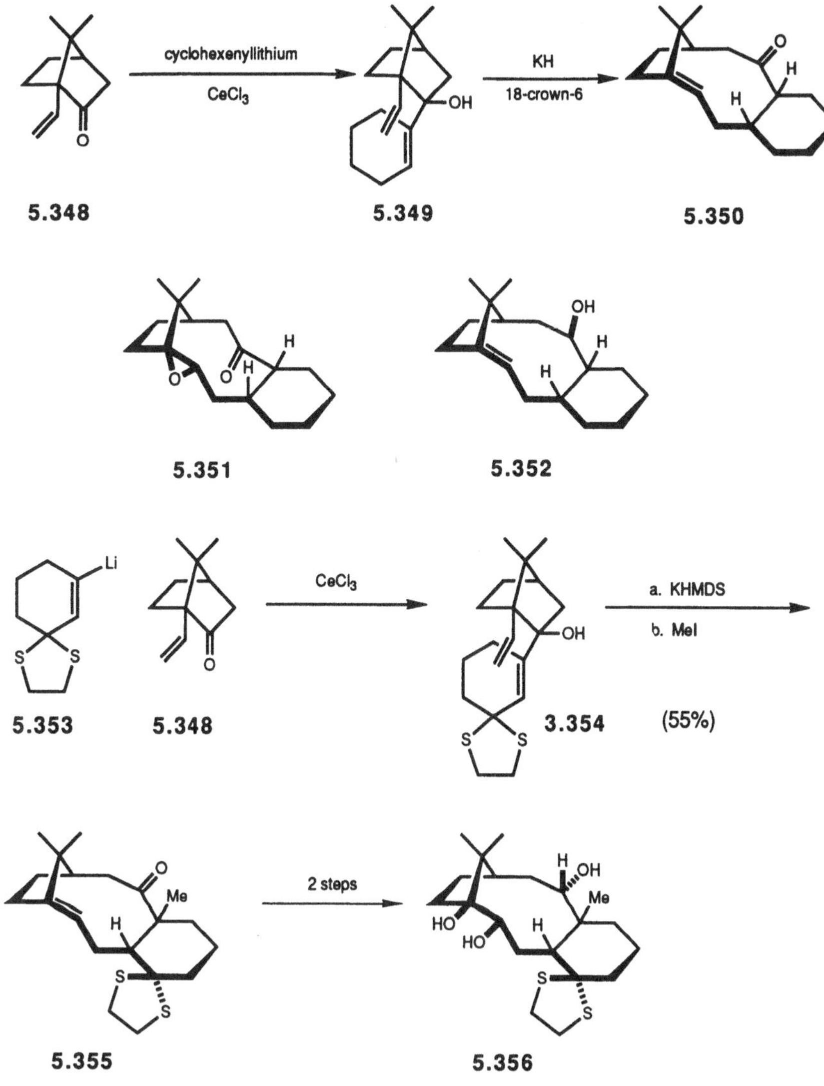

Scheme 5.66. Paquette's preparation of tricycles via the oxy-Cope rearrangement

with the optically pure bicyclic ketone (5.348). Reaction with cyclohexen-yllithium produced (5.349) which was then subjected to fragmentation conditions to produce (5.350). This intermediate was then epoxidized to create (5.351) or reduced to yield (5.352). The second sequence produced the rearrangement precursor (5.354) by coupling the functionalized cyclohexenyllithium (5.353) with (5.348). Upon rearrangement, (5.355) is produced in 55% yield. The two final steps transforms (**5.355**) into triol (**5.356**).

The most recent publication from PAQUETTE's group (*66*) describes the conversion of tricycles similar to (**5.348**) into the taxoid skeleton. The initial test is shown in Scheme 5.67. Enantiomerically pure (**5.357**), available in two steps form D-2-oxo-7,7-dimethyl-1-vinylbicyclo [2.2.1] heptane, was converted to diol (**5.359**) via intermediate (**5.358**). Mesylation followed by Wagner -Meerwein rearrangement led to the tricyclic taxoid model (**5.360**).

Since a number of taxanes possess a bridgehead hydroxyl, PAQUETTE *et al.* investigated a route designed to incorporate this C-1 hydroxyl (Scheme 5.68). The key rearrangement intermediate (**5.363**) was prepared from (**5.361**) in four steps via (**5.362**). Lewis acid catalyzed rearrangement resulted in the C-1 hydroxyl taxoid model (**5.364**) in 90% yield.

The successful production of (**5.360**) and (**5.364**) established the viability of this type of rearrangement in creating the tricyclo[9.3.1.0^{3,8}]-pentadecane ring system, but in order for this methodology to serve as the basis for an enantiospecific synthesis of the taxoids the correct trans B/C stereochemistry would have to be established. To this end, the conversion of (**5.357**) to (**5.365**) was investigated. This was readily achieved using the two step sequence shown in Scheme 5.69. Application of this same two step sequence to (**5.361**) led to a 1:1 mixture of (**5.366**) and (**5.367**). The efficiency of this process could be improved by recycling (**5.366**) after chromatographic separation.

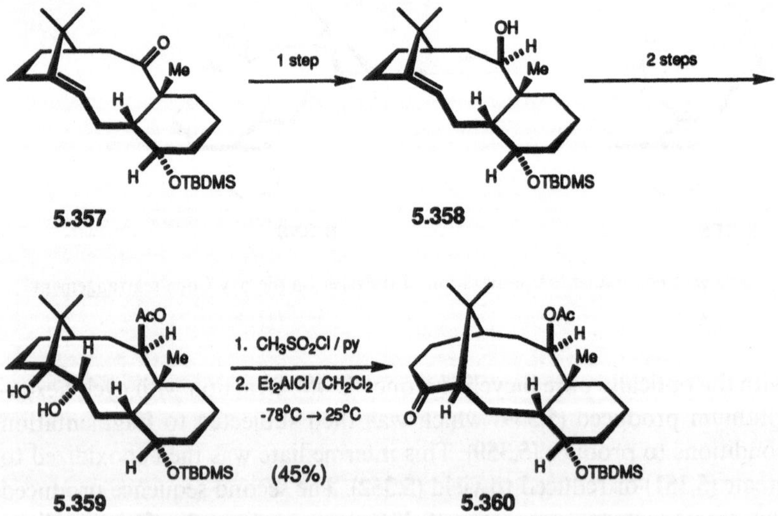

Scheme 5.67. Paquette's preparation of the taxoid skeleton

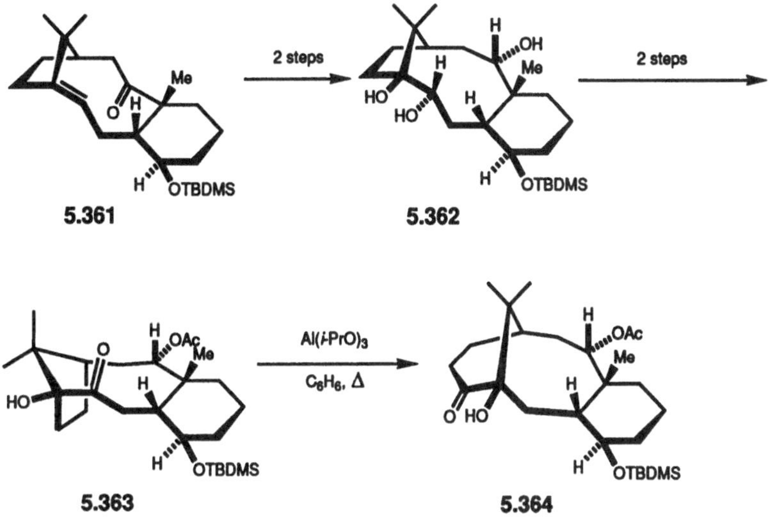

Scheme 5.68. Paquette's preparation of the taxoid skeleton with a C-1 hydroxyl

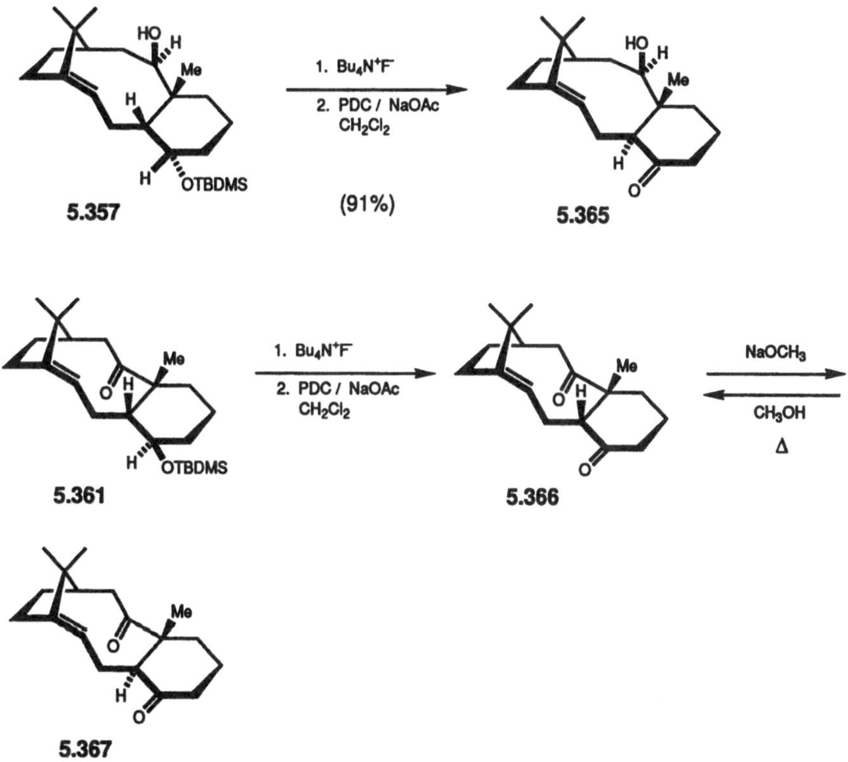

Scheme 5.69. Paquette's methodology for creation of the *trans*-fused BC stereochemistry

5.2.2.11. Snider's Approach

SNYDER and ALLENTOFF recently disclosed (*250*) the results of an extensive investigation that relies on the oxy-Cope rearrangement of 1,2-divinylcyclobutanols to create the AB portion of the taxoid skeleton. The initial and simplest implementation of this strategy is outlined in Scheme 5.70. Alkylation of (**5.368**) with (**5.369**) or (**5.370**) followed by intramolecular ketene [2 + 2] cycloaddition chemistry led to the desired cyclobutanones (**5.371**) and (**5.372**). Treatment of (**5.371**) and (**5.372**) with vinyllith-

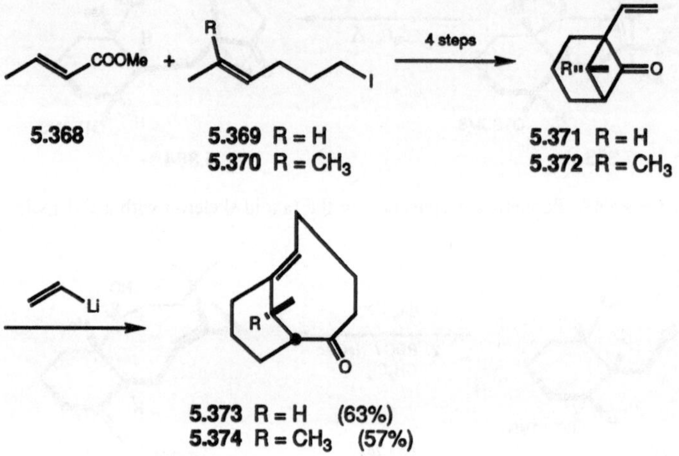

Scheme 5.70. Snider's preparation of AB taxoid models

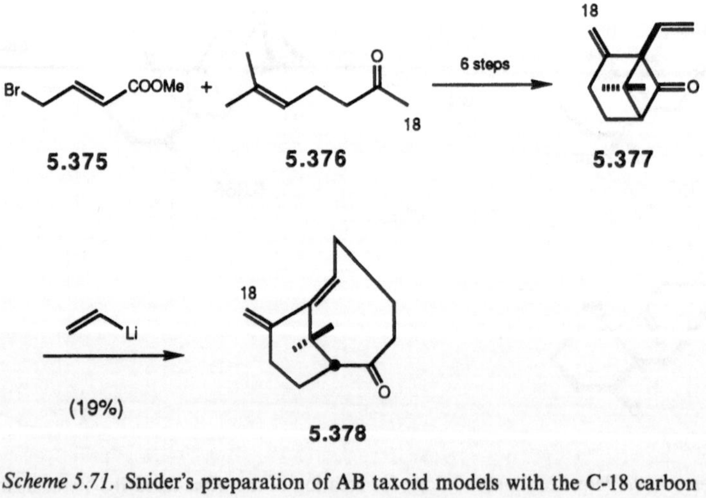

Scheme 5.71. Snider's preparation of AB taxoid models with the C-18 carbon

References, pp. 173–189

ium resulted in addition and then spontaneous rearrangement to yield the E-bicyclo[5.3.1]undecenones (**5.373**) and (**5.374**).

Once the oxy-Cope rearrangement of (**5.371**) and (**5.372**) had been demonstrated, they turned his attention to a route that would allow the incorporation of the taxoid C-18 methyl group (Scheme 5.71). In this sequence the desired vinyl cyclobutanone (**5.377**) was prepared from methyl 4-bromocrotonate (**5.375**) and 6-methyl-5-hepten-2-one (**5.376**) in six steps. Exposure to vinyllithium led to rearrangement to give the bicyclo[5.3.1]undecenone (**5.378**) in 19% yield.

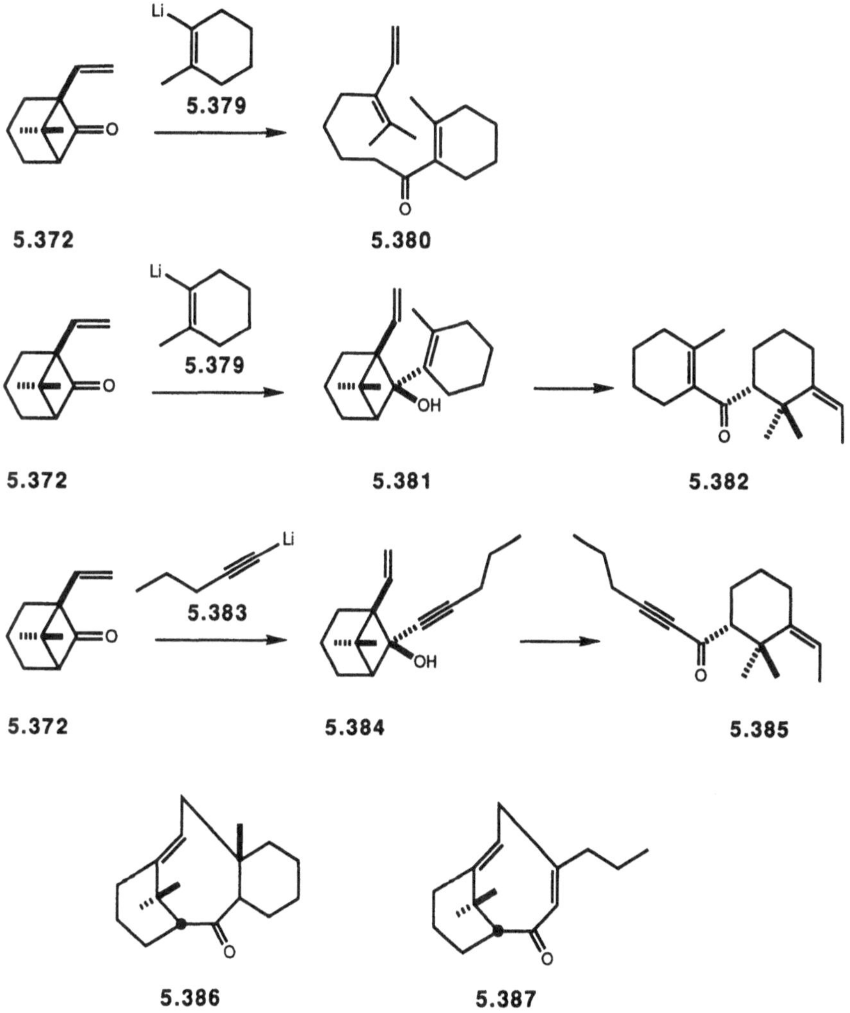

Scheme 5.72. Snider's attempt to construct the tricyclic taxoid skeleton

Snider originally envisioned the addition of a more highly substituted alkenyllithium reagent that would allow the construction of the tricyclic taxoid skeleton by a convergent process. The results of efforts directed toward this goal are shown in Scheme 5.72. Treatment of bicycloheptanone (**5.372**) with cyclohexenyllithium (**5.379**) led to (**5.380**). Addition of (**5.372**) to lithium reagent (**5.379**) followed by an acid quench produced (**5.381**). Attempts to induce the thermal oxy-Cope rearrangement of (**5.381**) resulted in the retro-ene product (**5.382**) and not the desired tricycle (**5.386**). A third attempt involved the reaction of 1-pentynyllithium (**5.383**) with (**5.372**). As was previously observed, this resulted in the addition product (**5.384**), but this failed to undergo the desired rearrangement. Attempts to induce thermal oxy-Cope rearrangement again led to the retro-ene product (**5.385**) rather than the oxy-Cope product (**5.387**).

5.2.2.11. Zucker's Approach

A 1990 (*316*) by Zucker and Lupia describes an oxy-Cope rearrangement approach that is very similar to the one investigated by Snider. As

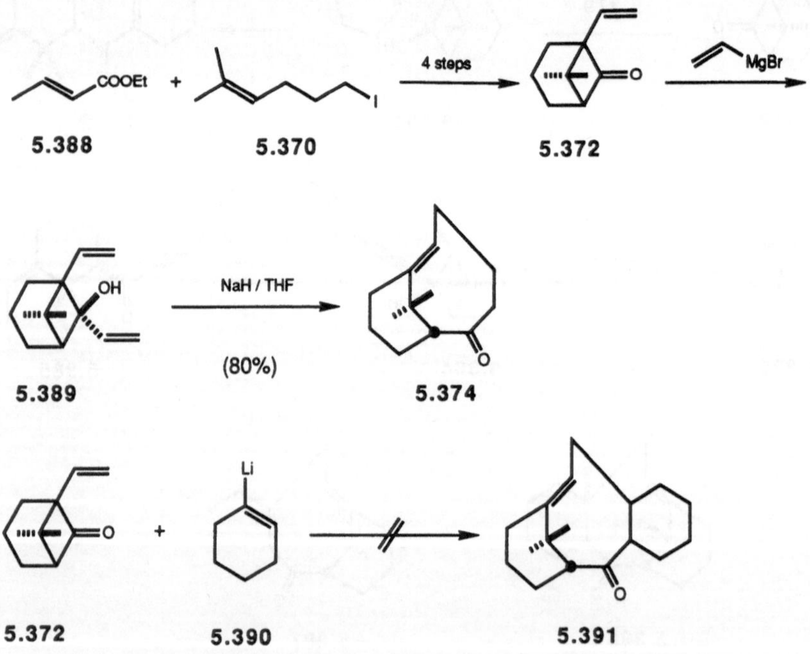

Scheme 5.73. Zucker's anionic oxy-Cope approach to the taxoid skeleton

in the preceding section these workers envisioned the construction of the taxoid skeleton using the oxy-Cope reaction of a suitably substituted divinyl cyclobutanol that would lead directly to the taxoid skeleton upon rearrangement. The results of this investigation are outlined in Scheme 5.73.

The initial target was the AB model (**5.374**). The requisite ketone (**5.372**) was prepared in four steps from ethyl crotonate (**5.388**) and iodide (**5.370**). Addition of vinyl magnesium bromide led to intermediate (**5.389**) which set the stage for the oxy-Cope rearrangement. When (**5.389**) was treated with NaH, rearrangement occurred to give the desired bicyclic keto-olefin (**5.374**). Attempts to introduce what would be the C-18 methyl group of the taxoids using this route proved to be unsuccessful.

Since ZUCKER and LUPIA envisioned the preparation of the taxoid skeleton by application of this methodology to a convergent approach, they briefly examined the reaction of cyclohexenyllithium (**5.390**) with ketone (**5.372**). This led to the formation of a mixture of product from which none of the desired product, (**5.391**), could be isolated.

5.3. Partial Synthesis of Taxol and Related Compounds

The major goal of synthetic efforts in the synthesis of taxane diterpenoids is of course taxol because of its important antineoplastic activity. Although taxol has not yet yielded to total synthesis, several synthetic approaches have appeared that start with the taxoid baccatin III (**1.10**) or its 10-deacetyl analog. Since 10-deacetylbaccatin III can be obtained in yields of up to 0.1% from *T. baccata* leaves (*55*) a successful synthesis of taxol from this material might alleviate the supply problems associated with the drug.

5.3.1. Synthesis of Taxol and Taxol Analogs from 13-Cinnamoylbaccatin III

The first successful partial synthesis of taxol was developed by POTIER and his collaborators through reaction of 13-cinnamoylbaccatin III with various reagents. This approach was adopted initially because baccatin III and its 10-deacetyl analog are very resistant to esterification at C-13 as noted earlier (Sect. 4.3.1) and only the unhindered cinnamoyl group could be introduced readily.

Initial studies were carried out on 10-deacetylbaccatin III (**5.392**) (*236*) (Scheme 5.74). Treatment of this compound with trichloroethyl

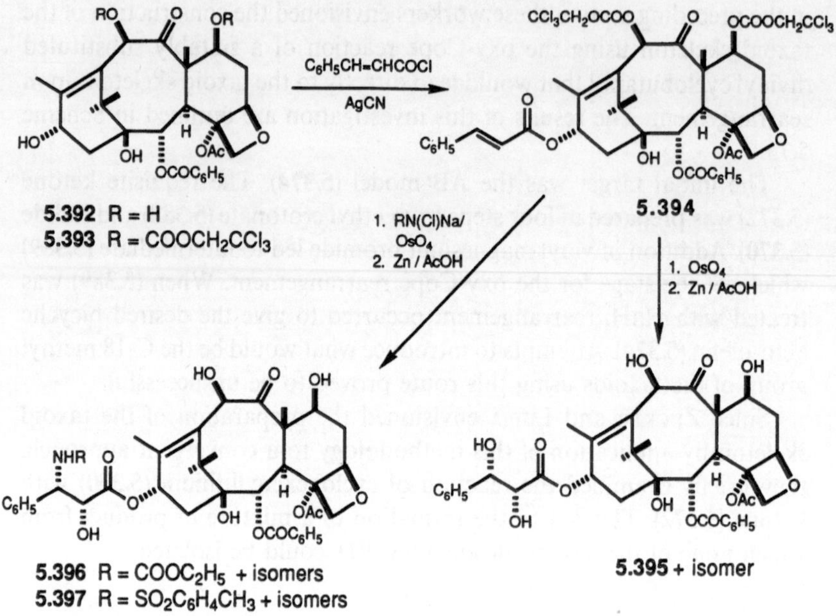

5.392 R = H
5.393 R = COOCH2CCl3
C6H5CH=CHCOCl
AgCN
5.394
1. RN(Cl)Na/
OsO4
2. Zn / AcOH
1. OsO4
2. Zn / AcOH
5.396 R = COOC2H5 + isomers
5.397 R = SO2C6H4CH3 + isomers
5.395 + isomer

Scheme 5.74. Synthesis of 10-deacetyltaxol analogs from 10-deacetyl baccatin III

chloroformate gave the 7,10-protected derivative (**5.393**) and reaction with cinnamoyl chloride in the presence of silver cyanide gave the cinnamoyl ester (**5.394**). Reaction of (**5.394**) with osmium tetroxide followed by deprotection then yielded the diol mixture (**5.395**), while reaction under Sharpless hydroxyamination conditions yielded the protected amino alcohols (**5.396**) and (**5.397**). In each case the products shown were formed together with the diastereomeric products resulting from attack of the oxidant from the opposite face of the double bond, and in the case of the formation of (**5.396**) and (**5.387**) the regioisomers with the hydroxyl and amino groups reversed were also formed.

A later publication from the same group (*42*, *172*) gave full details of the chemistry described above and carried it through to prepare taxol itself. Application of the Sharpless hydroxyamination procedure to the cinnamate ester (**5.394**) using N-chloro-N-sodio-tert-butylcarbamate as the amine component gave the N-t-BOC derivative (**5.398**) together with its regio- and diastereo-isomers. The yield of the desired diastereomers could be improved by carrying out the reaction in the presence of dihydroquinine acetate or chlorobenzoate; under the best conditions (**5.398**) and its regioisomer with the amino and hydroxyl groups reversed were obtained as an approximately 1:1 mixture in 67% yield. Conversion of (**5.398**) to 10-deacetyltaxol (**5.399**) was achieved by cleavage of the

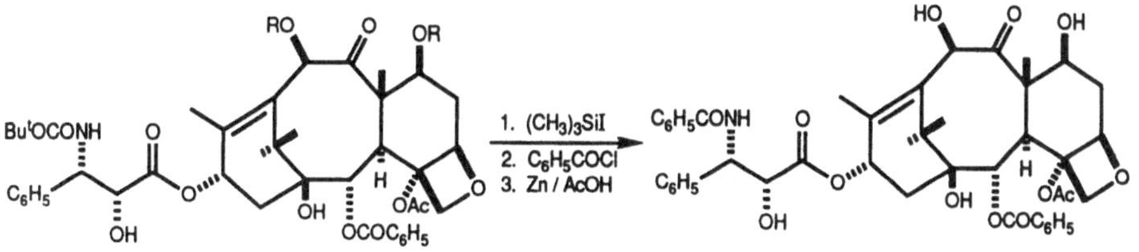

Scheme 5.75. Synthesis of 10-deacetyltaxol

N-t-BOC group with trimethylsilyl iodide, benzoylation, and removal of the trichloroethyl carbonyl protecting groups with zinc in acetic acid (Scheme 5.75). A similar overall sequence applied to baccatin III as starting material gave taxol as the final product; the yield of taxol from 7-troc-baccatin III was about 14%.

The chemistry described above has obvious limitations, chief among them being the almost complete lack of regiochemical control and the partial lack of stereochemical control in the oxyamination reaction. Nevertheless, this work did have an important benefit, since it resulted in the synthesis of N-debenzoyl-N-t-butoxycarbonyl-10-deacetyltaxol (5.400) by deprotection of intermediate (5.398). This compound, now named RP56976 or taxotere, has been shown to be somewhat more active than taxol in certain assays and it is currently in clinical trials as an antineoplastic agent. In addition, the formation of several isomers enabled POTIER and his collaborators to prepare and test a large number of taxol and 10-deacetyltaxol analogues, and thus provide valuable data for establishing structure-activity relationships (88).

5.3.2. Synthesis of Taxol by Acylation of Baccatin III with a Pre-formed Side Chain.

The problem of acylating baccatin III with a suitably protected C-13 side chain was first solved by GREEN and POTIER (55). These investigators treated 7-(triethylsilyl)baccatin III (5.401) with excess (2R, 3S)-N-benzoyl-O-(1-ethoxyethyl)-3-phenyl isoserine (5.402) in the presence of di-2-pyridyl carbonate (DPC) and 4-dimethylaminopyridine (DMAP) at 73° in toluene for 100 h and obtained the protected taxol derivative (5.403) in 80% yield at 50% conversion (Scheme 5.76). Removal of the protecting groups with dilute HCl then gave taxol in good yield. These conditions were developed after a great deal of experimentation by their

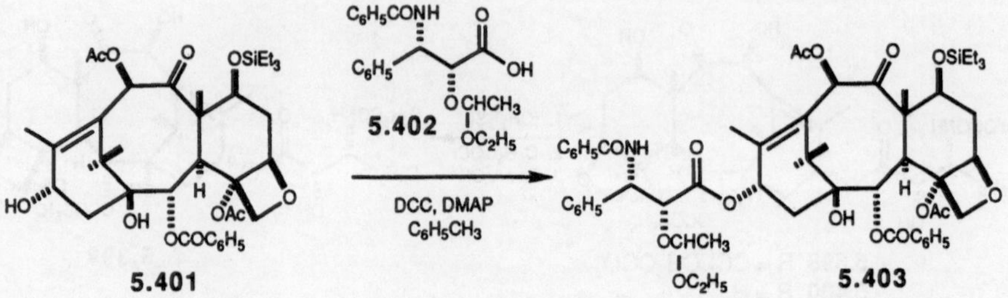

Scheme 5.76. Synthesis of taxol by direct acylation of baccatin III with a preformed side chain

group and other groups and are very specific; many other conditions have been tried and have failed to yield significant amounts of taxol.

The synthesis of *threo*-3-phenylisoserine has been accomplished by several groups. Early syntheses gave mixtures of *threo* and *erythro* isomers (*125*) or racemic material. Thus racemic *threo*-phenylisoserine was prepared in good yield from commercially available racemic *threo*-phenylserine (**5.404**) by reaction with nitrous acid in the presence of potassium bromide to give the α-bromo-β-hydroxycarboxylic acid (**5.405**). This acid was then treated with concentrated aqueous ammonia to give the desired product (**5.406**) together with a small amount of its regioisomer (Scheme 5.77) (*44*). Reaction was shown to proceed through *cis*-β-phenylglycidic acid, and ammonolysis of ethyl *cis*-β-phenylglycid-ate was shown much earlier to yield predominantly *threo*-phenylisoserin-amide (*126*). Since *cis*-β-phenylglycidic acid has been resolved (*60, 93*), this methodology could in principle be used to prepare optically active material. An alternate strategy was selected by HÖNIG and coworkers who resolved the 2-butanoyl ester of *threo*-3-azido-2-hydroxy-3-phenylpropionic acid with lipase from *Pseudomonas fluorescens*. The unhydrolysed ester, obtained in 35% yield and 98% enantiomeric excess, had the correct absolute configuration for the taxol side chain, and could be hydrolysed and hydrogenated to (*2R, 3S*)-3-phenylisoserine (*110*).

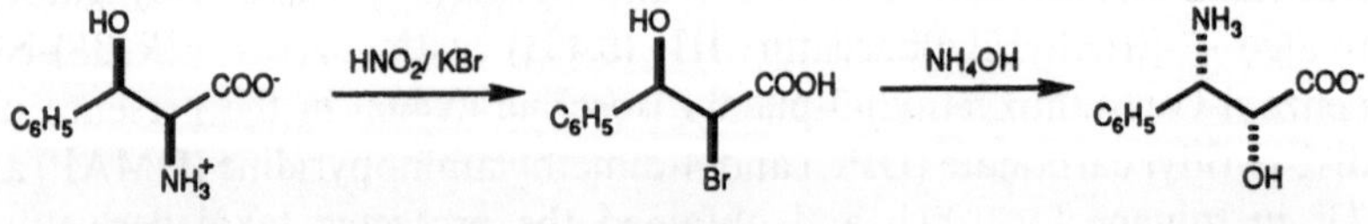

Scheme 5.77. Synthesis of racemic *threo*-3-phenylisoserine from racemic *threo*-phenylse-rine

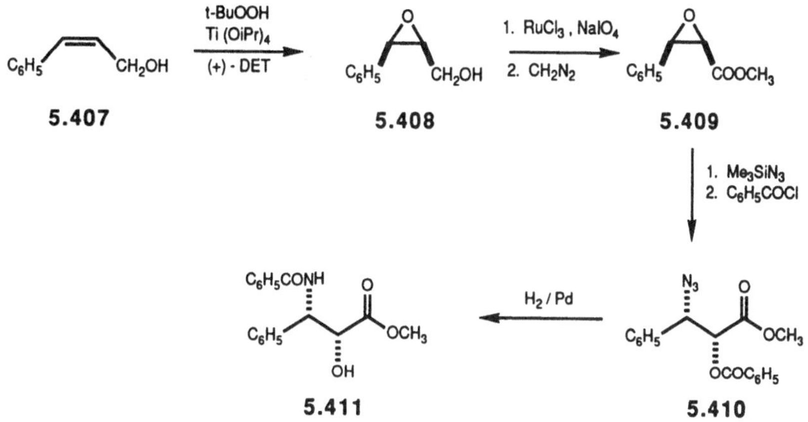

Scheme 5.78. Synthesis of scalemic N-benzoyl-*threo*-3-phenylisoserine by Sharpless methodology

The first enantioselective synthesis of the taxol side chain was developed by GREENE and his collaborators (Scheme 5.78) (*56*). These workers used Sharpless epoxidation chemistry to convert *cis*-cinnamyl alcohol (**5.407**) to the epoxide (**5.408**) in a yield of 61% with 78% enantiomeric excess. Oxidation of (**5.408**) and immediate methylation gave the methyl glycidate (**5.409**), which underwent regio-and stereoselective ring opening to an azido alcohol which was benzoylated to give (**5.410**). Hydrogenation of (**5.410**) was accompanied by acyl migration to give the desired ester (**5.411**) in 23% overall yield.

An improved synthesis of the key glycidic ester (**5.409**) was later developed by the same group (Scheme 5.79) (*53*). Oxidation of the readily available methyl cinnamate (**5.412**) with osmium tetroxide by the Sharpless procedure in the presence of N-methylmorpholine N-oxide (NMMO) and dihydroquinidine 4-chlorobenzoate (DQCB) gave the diol (**5.413**) which could be recrystallized to enantiomeric purity in 51% yield. Monotosylation of this diol gave only the C-2 tosylate, which then gave glycidic ester (**5.409**) on treatment with potassium carbonate in 35% overall yield from methyl cinnamate.

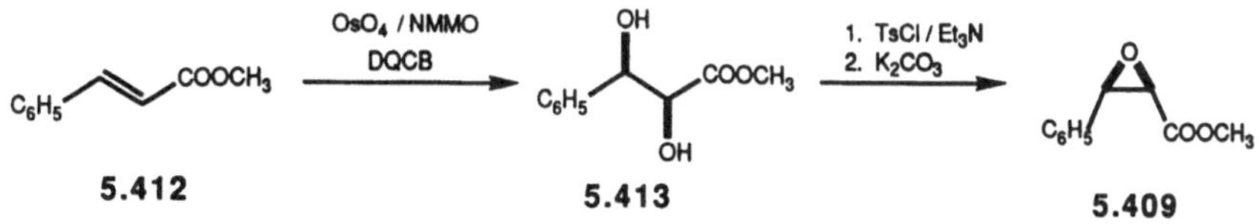

Scheme 5.79. Improved synthesis of key glycidic ester **5.409**

A recent publication from the GREENE group provides a third route to the taxol and taxotere side chains (Scheme 5.80) (*54*). In this approach the readily available (S)-(+)-phenylglycine (**5.414**) is reduced and benzoylated to give the amido alcohol (**5.415**). Oxidation of (**5.415**) to the aldehyde by the Swern procedure sets the stage for an enantioselective Grignard reaction to give the alcohol (**5.416**), which was protected with ethyl vinyl ether and oxidized to the protected acid (**5.417**). The overall yield of chemically and enantiomerically pure acid was 30%; a corresponding process gave the taxotere side chain in 34% yield.

The taxol side chain has also been prepared by various pathways involving a β-lactam intermediate. In one of these, the racemic *cis*-β-lactam (**5.421**) or (**5.422**) is prepared by reaction of the alkoxyacylchloride (**5.418**) or acetyloxyacyl chloride (**5.419**) with imine (**5.420**) (Scheme 5.81). Removal of the *p*-methoxyphenyl (PMP) protecting group of (**5.421**) with ceric ammonium nitrate (CAN), followed by ring opening and benzoylation, gave the racemic methyl ether of the side chain (**5.423**) (*205*). Appropriate protective group manipulation would presumably allow the synthesis of the side chain with a different ether protective group, and

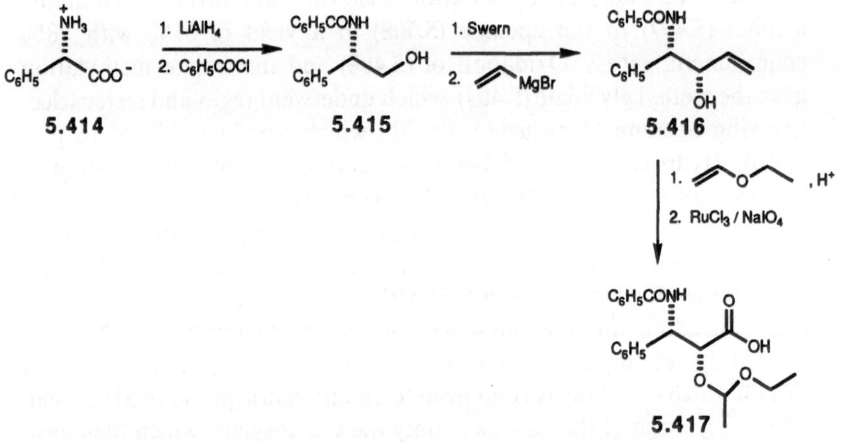

Scheme 5.80. Improved synthesis of protected acid **5.417**

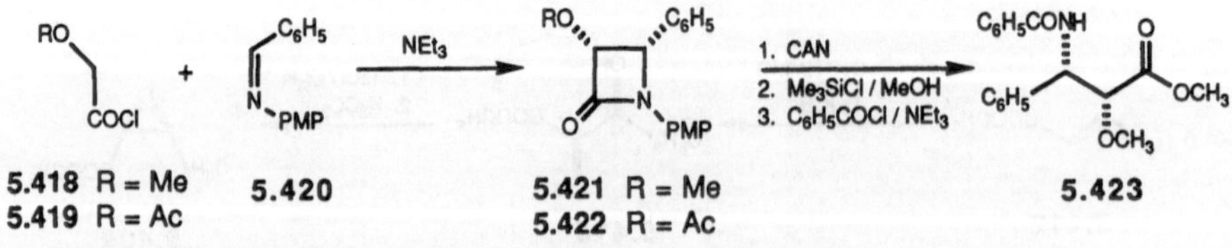

Scheme 5.81. Synthesis of racemic side chain by a β-lactam pathway

References, pp. 173–189

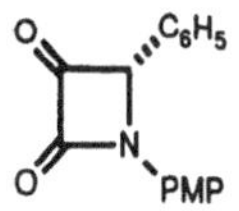

5.424

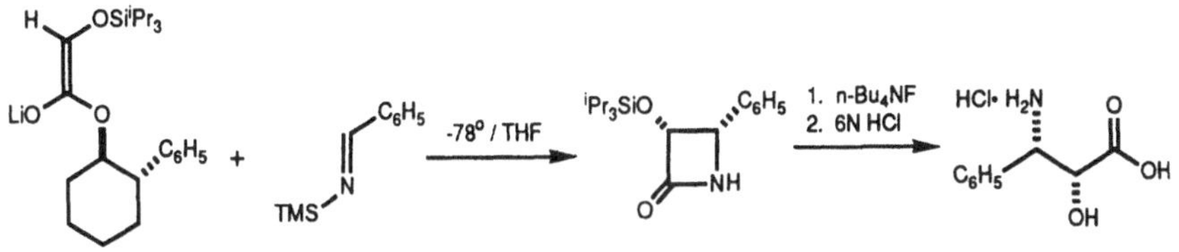

| **5.425** | **5.426** | **5.427** | **5.428** |

Scheme 5.82. Synthesis of scalemic side chain by a β-lactam pathway

these authors did indeed develop a synthesis of the taxol side chain from the azetidine-2,3-dione (**5.424**).

An asymmetric synthesis of the side chain using β-lactam chemistry has been developed by OJIMA and his collaborators (*204*). In this chemistry, (Scheme 5.82), the lithium enolate (**5.425**) is condensed with the N-TMS-imine (**5.426**) to give the β-lactam (**5.427**) with 96–98% enantiomeric excess. Deprotection of (**5.427**) with fluoride ion followed by hydrolysis yields the amino acid (**5.428**), which can be benzoylated to give the side chain (**5.411**).

5.3.3. *Acylation of Baccatin III with β-Lactams or Oxazinones*

In addition to the acylation of baccatin III by a protected N-benzoyl-3-phenyl isoserine such as (**5.402**), acylation has also been achieved with β-lactams or oxazinones as the acylating agent. Thus reaction of 7-(triethylsilyl)baccatin III with excess racemic N-benzoyl β-lactam (**5.429**) in the presence of DMAP and pyridine at 25°, followed by deprotection with dilute acid, gave taxol in good yield (*107*). Similar results were obtained with oxazinones; thus reaction of excess racemic oxazinone (**5.430**) with 7-(triethylsilyl) baccatin III in pyridine and DMAP at 25° gave a 77% yield of 2'-(1-ethoxyethyl)-7-(triethylsilyl)taxol, which was hydrolysed to taxol in 90% yield (*108*). It is worth noting that SWINDELL has proposed that the direct acylation of baccatin III with protected N-benzoyl-3-phenylisoserines proceeds through oxazinone intermediates, on the basis of the fact that the rate of

5.429

5.430

reaction is actually enhanced by the presence of the N-benzoyl group (*263*).

The union of the HOLTON β-lactam coupling method with the OJIMA enantioselective lactam synthesis described earlier provides an excellent route to the synthesis of taxol. This chemistry is thus very competitive with the POTIER-GREENE approach described in the previous section. The yields in the key coupling step appear to be better in HOLTON'S approach, but the synthesis of the chiral acid appears to be most facile by the third GREENE approach described above.

5.3.4. Synthesis of Taxol Analogues

A number of taxol analogues have been prepared by partial synthesis. Two groups have prepared taxols modified in the side chain by coupling appropriate side chains to baccatin III. SWINDELL and his collaborators prepared analogues lacking various side chain groups by coupling of the appropriate acid to a protected baccatin III. In this way they prepared side chains lacking the 2′-OH group, the 3′-phenyl group, or the 3′-N-benzoyl group (*263*). POTIER and his collaborators used the cinnamoyl-ation procedure described earlier to prepare a large number of regio-and stereoisomers at the 2′- and 3′-positions of the taxol side chain (*88*). A discussion of the structure-activity relationships that emerge from these and other studies is given in a recent review (*131*).

Three groups have prepared analogs with an intact side chain but a highly modified "taxane" nucleus. PIRRUNG and DEAMICIS coupled race-mic taxol side chain with three different diiodoalkanes to prepare the symmetrical diesters (**5.431–5.433**); none of these showed any activity (*44*). FILLION has prepared the naphthalene carbolactone ester (**5.434**), but no biological data have been presented for this compound (*75*). Finally, BLECHERT and KLEINE-KLAUSING have prepared the analogue (**5.435**) by acylation of taxoid model (**5.275**) (Sect. 5.2.2.3), and reported that it shows tubulin assembly activity (*18*). Unfortunately quantitative data were not reported, so it is difficult to judge the significance of this result: if the activity is anywhere close to that of taxol, however, this

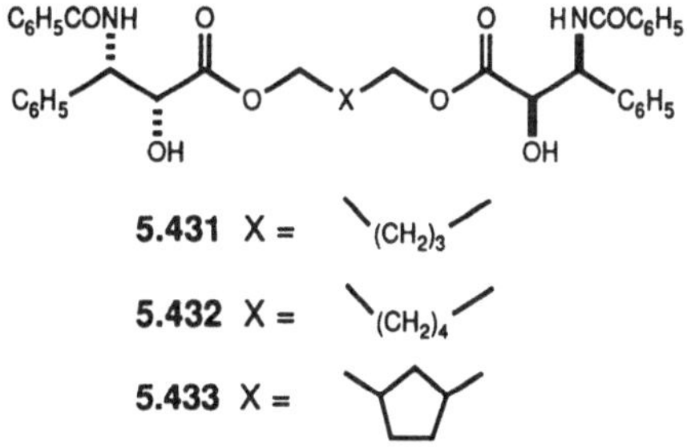

5.431 X = $(CH_2)_3$

5.432 X = $(CH_2)_4$

5.433 X =

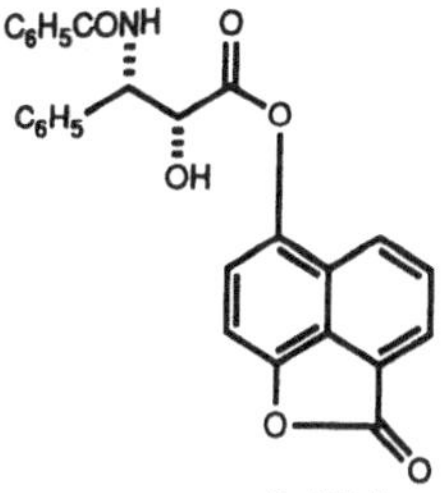

5.434

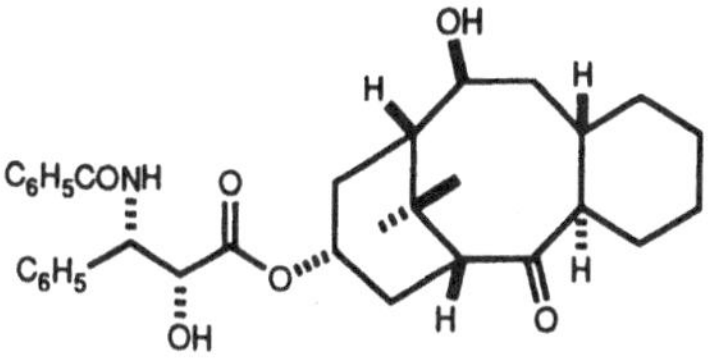

5.435

result would be very important indeed, since it would indicate that an oxetane ring is not necessary for biological activity of taxol analogues.

5.3.5. Synthesis of Oxetane Models

The oxetane ring of taxol is a unique structural feature of the molecule which, in spite of the result just noted, is believed to be necessary for its biological activity (230) and has been the subject of some independent synthetic studies. In the first approach, (Scheme 5.83) cyclohexene (5.436) was epoxidized to epoxy acetate (5.437), and treatment of this compound with $BF_3 \cdot Et_2O$ gave the diol (5.438), presumably through an acetoxonium ion intermediate. Closure of the diol to the

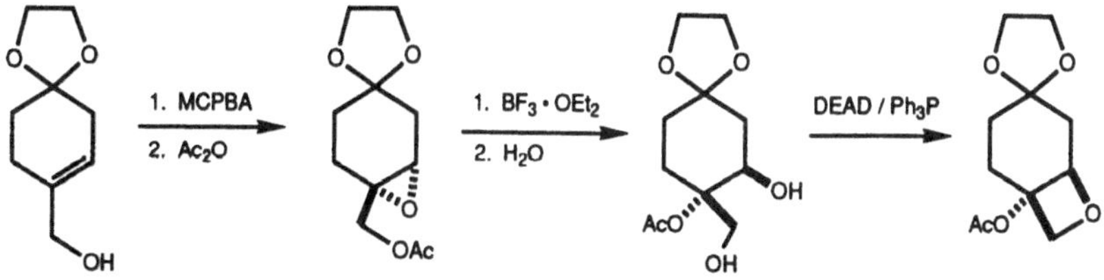

5.436   **5.437**   **5.438**   **5.439**

Scheme 5.83. Berkowitz' oxetane synthesis

oxetane (**5.439**) was then accomplished by a modification of the Mitsun-obu procedure (*12*).

A second approach examined the possible intermediary of an epoxy alcohol such as (**5.440**) in the biogenesis of the oxetane ring. As described more fully in section 6.1, solvolysis of mesylate (**5.441**) failed to yield oxetane (**5.442**) (*256*).

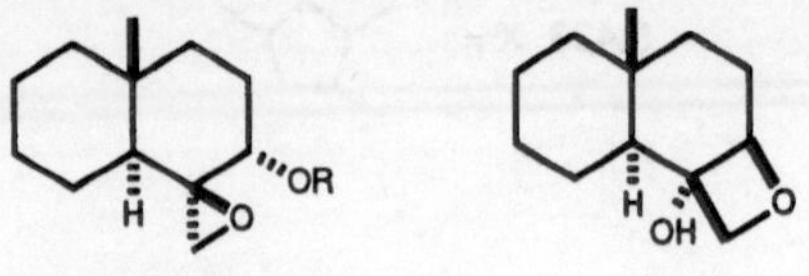

5.440 R = H
5.441 R = SO₂CH₃

5.442

A third approach has some parallels to the first one above (Scheme 5.84). Hydroxylation of diene (**5.443**) gave the diol (**5.444**) as an insepar-able mixture of diastereomers together with its regioisomer. Mesylation of the primary hydroxyl group and protection of the tertiary hydroxyl group gave the mesylate (**5.445**) which was converted to oxetane (**5.446**) by tetrabutylammonium fluoride in refluxing THF. The yield of (**5.446**) from the 3-methyl-2-cyclohexen-1-one starting material was 5%. Cycliz-ation of mesylate to (**5.446**) failed if the tertiary hydroxyl group was not protected, presumably because deprotection of the tertiary hydroxyl by fluoride ion led to rearrangement to an epoxide analogous to (**5.440**) (*160*).

Very recently a related strategy has been applied to attachment of an oxetane ring to a taxoid skeleton (Scheme 5.85) (*71*). Taxine B (**1.6**) was converted to 5-O-cinnamoyl-taxicin I triacetate (**1.2**) by previously de-scribed chemistry and thence to the diacetonide derivative (**5.447**). Hy-droxylation of (**5.447**) with osmium tetroxide in the presence of N-methylmorpholine N-oxide gave the diol (**5.448**) in 81% yield. Diol (**5.448**) was converted to mesylate (**5.449**) by protection of the primary hydroxyl group as its t-butyldimethylsilyl ether, mesylation, and depro-tection. Treatment of (**5.447**) with strong bases such as sodium hydride or

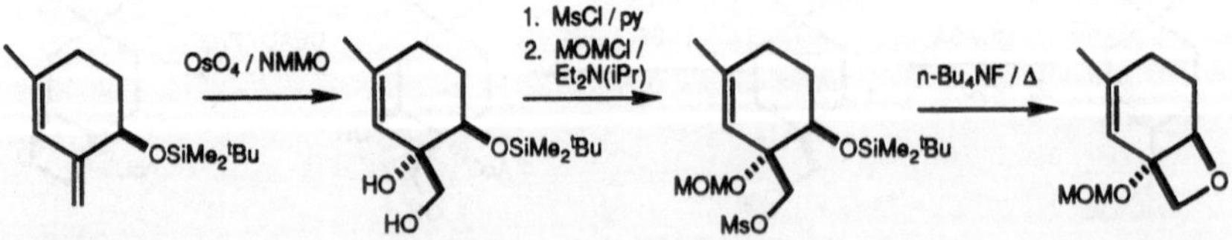

5.443   **5.444**   **5.445**   **5.446**

Scheme 5.84. Clark's oxetane synthesis

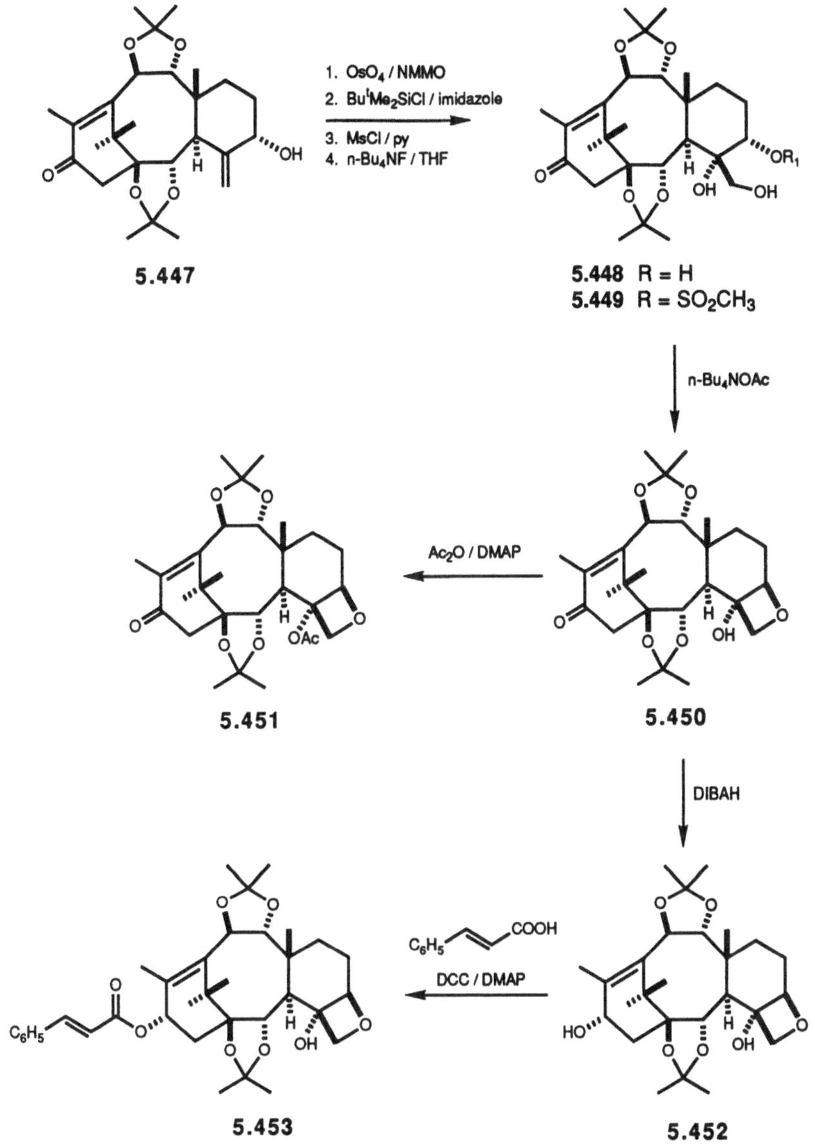

Scheme 5.85. Synthesis of taxoids with oxetane rings

potassium t-butoxide led to formation of various 4-oxo products, but treatment with the weak base ammonium acetate gave oxetane (**5.450**) in good yield. Oxetane (**5.450**) could be acetylated with difficulty to the acetate (**5.451**), or reduced with DiBAH to the 13α alcohol (**5.452**) in 39% yield (other reducing agents tested gave either conjugate addition or gave mainly the 13β-alcohol). Cinnamoylation of (5.452) gave the cinnamate ester (5.453), which is obviously convertible into taxol analogs by methods similar to those described earlier. This work thus makes a significant advance in the preparation of taxol analogs from other readily available taxoids.

6. Biosynthesis and biotransformation of taxoids

6.1. Biosynthesis of taxoids

No studies dealing with the biosynthesis of taxol or any other taxoids have been published so far, but some work has been carried out on the biosynthesis of Winterstein's acid. As noted earlier, acid hydrolysis of taxine yields (3R)-dimethylamino-3-phenylpropanoic acid (Winterstein's acid, 6.2) (298, 299). Biosynthetic studies on this acid have been carried out by LEETE and BODEM (157) and by HASLAM (212). In T. baccata phenylalanine (6.1) serves as the best precursor of this acid, and the biosynthetic conversion occurs stereospecifically, with retention of the 3-pro-S proton (Scheme 6.1).

The taxane ring system itself most probably arises by electrophilic cyclization of geranyl geranyl pyrophosphate (6.3), possibly through a cationic intermediate such as (6.4). The formation of epi-verticillol (6.5) and/or epi-cembrene (6.6) intermediates has been suggested (8, 97) (Scheme 6.2). It should be noted that the published absolute configuration of verticillol is the 1-epimer of that shown (taxane numbering) (127), and thus any taxane formed from verticillol itself would be epimeric at C-1 with the naturally occurring compounds.

Although the proposed biosynthetic pathway is eminently reasonable, experimental support has so far not been achieved. No incorporation studies on taxol or any other taxane have been reported and the model studies that have been carried out have not yielded products with a taxane skeleton. Thus treatment of cembrene (6.9) with acid yielded the tricyclic product (6.10) (Scheme 6.3) (43), while treatment of verticillene,

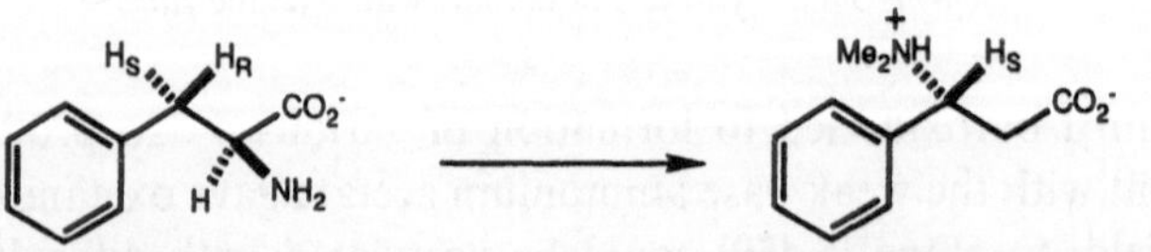

Scheme 6.1. Conversion of phenylalanine to Winterstein's acid in T. baccata

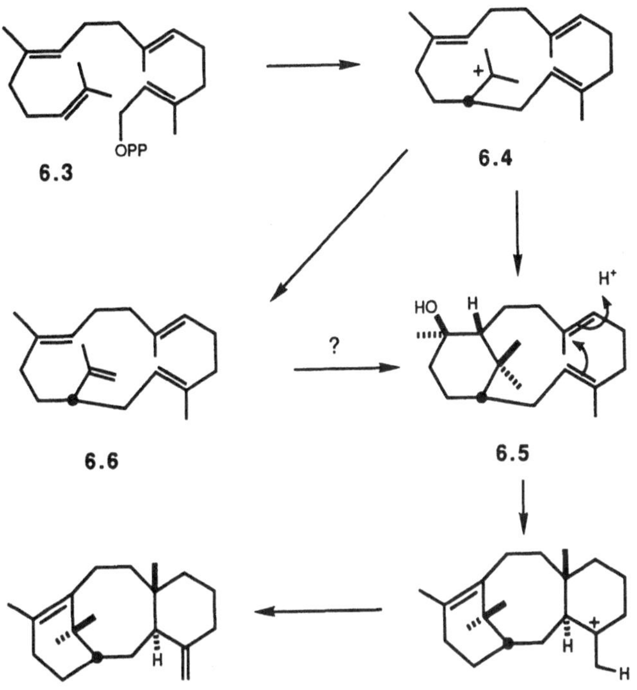

Scheme 6.2. Proposed biosynthesis of the taxene skeleton

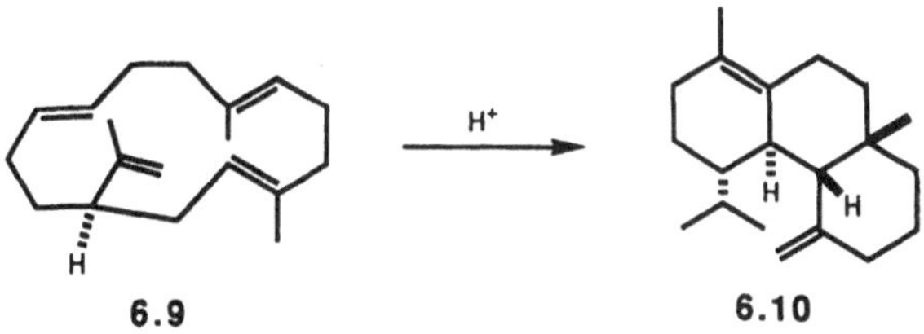

Scheme 6.3. Acid treatment of cembrene

verticillene 7,8-epoxide, or anhydro verticillol 7,8-epoxide with Lewis acids yielded only products resulting from rearrangements of the epoxide rings. Thus as one example verticillene 7,8-epoxide (6.11) yielded none of the expected cyclization product (6.12) on treatment with BF_3-etherate, instead forming the allylic alcohol (6.13) and its corresponding fluorohydrin (Scheme 6.4) (8). Although these experiments do not disprove the role of cembrene or verticillene-type intermediates in the biosynthesis of the taxoids, they do suggest that the biogenetic sequence is more subtle

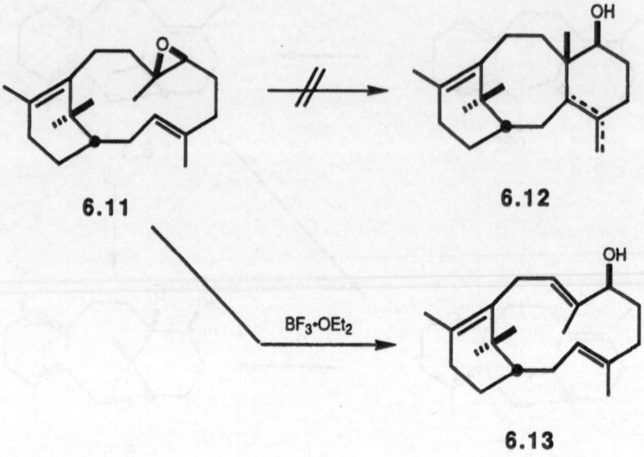

6.11   6.12

6.13

Scheme 6.4. Treatment of verticillene 7,8-epoxide with Lewis acids

than previously supposed. It could, for example, involve different geo-
metric or positional isomers of geranyl geranyl pyrophosphate or
perhaps alternate reactive intermediates such as radicals.

The one experiment that has yielded a skeleton at least related to the
taxane skeleton ironically was carried out without any avowed intention
of modeling the biosynthesis of the taxoids. Oxidation of the diterpenoid
cleomeolide **(6.14)** gave the ketoester **(6.15)**, which yielded lactone **(6.16)**
on treatment with mild base and the final product **(6.17)** on more
vigorous base treatment *(26)* (Scheme 6.5). Although **(6.16)** and **(6.17)** do

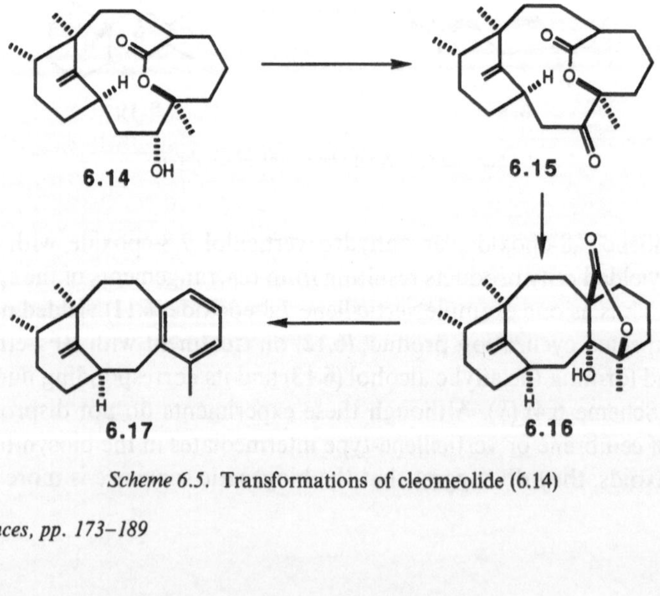

6.14 OH   6.15

6.17   6.16

Scheme 6.5. Transformations of cleomeolide (6.14)

not have the actual taxane skeleton, (**6.16**) differs only in the placement of a methyl group, and this transformation thus offers an interesting model for further work.

The important oxetane ring present in taxol and related taxoids has also been the subject of biosynthetic speculation and experiment. The first biosynthetic speculation by HALSALL and his collaborators (*47*) involved ring opening of an epoxide such as baccatin I (**6.18**) followed by recyclization of the intermediate (**6.19**) to form the oxetane (**6.20**) (Scheme 6.6); an unusual hydroxyl group migration was also proposed, but this is not an essential part of their scheme. An intermediate similar to (**6.19**) has been converted to an oxetane as previously described in Sect. 5.3.5 (*12*), so this part at least of the biosynthetic scheme is chemically reasonable.

A related but different proposal for the biosynthesis of the oxetane ring was made by SWINDELL and BRITCHER who suggested that solvolysis of a suitable epoxy alcohol could yield an oxetane through an intermediate oxabicyclobutonium ion (*256*). However, solvoylsis of the model

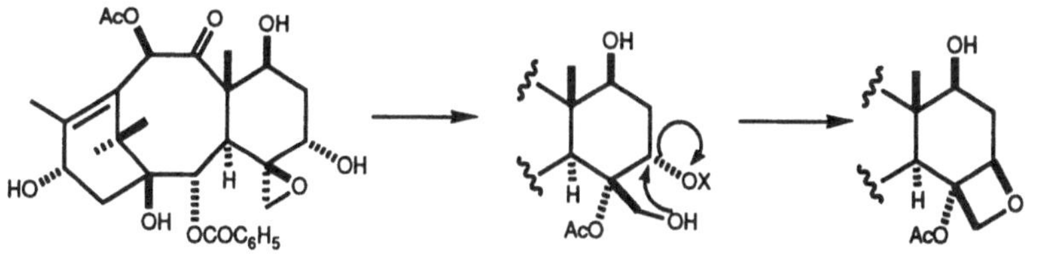

6.18   **6.19**   **6.20**

Scheme 6.6. Halsall's proposal for biogenesis of the oxetane ring

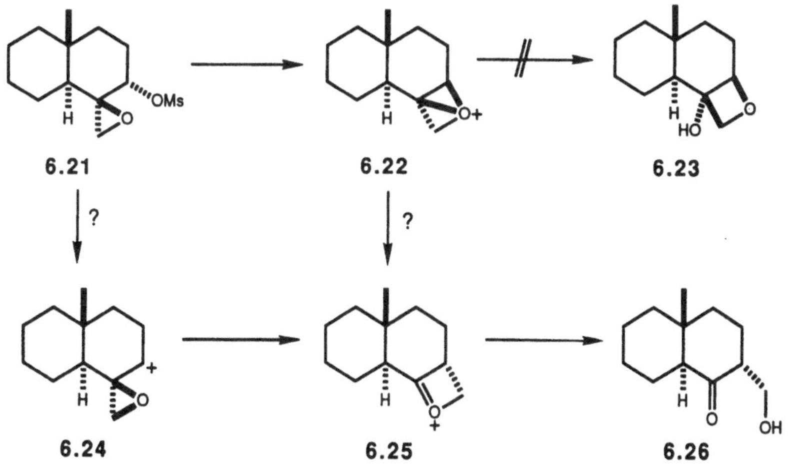

6.21   **6.22**   **6.23**

6.24   **6.25**   **6.26**

Scheme 6.7. Swindell's model studies on oxetane ring formation

compound (6.21) failed to yield any oxetane (6.23), instead giving only the ketoalcohol product (6.26). The intermediacy of the oxabicyclobutonium ion (6.22) is not excluded by this result since it could rearrange to the oxonium ion (6.25), but an alternate pathway to (6.25) via the cation (6.24) is perhaps more plausible (Scheme 6.7).

An alternative proposal for the biogenesis of the oxetane ring has been made by POTIER and his collaborators (87). He noted the existence of three major groups of taxanes: group A with an exocyclic methylene at C-4, group B with a β-epoxide (oxirane) at C-4, and group C with an oxetane ring. Many of the group A compounds are esterified at C-5 with acids (such as Winterstein's acid) that are structurally related to the taxol C-13 ester side chain. POTIER thus suggested that formation of the oxetane ring is preceded by an intramolecular transesterification from C-5 to C-13, as for example from (6.27) to (6.28) (Scheme 6.8). Although

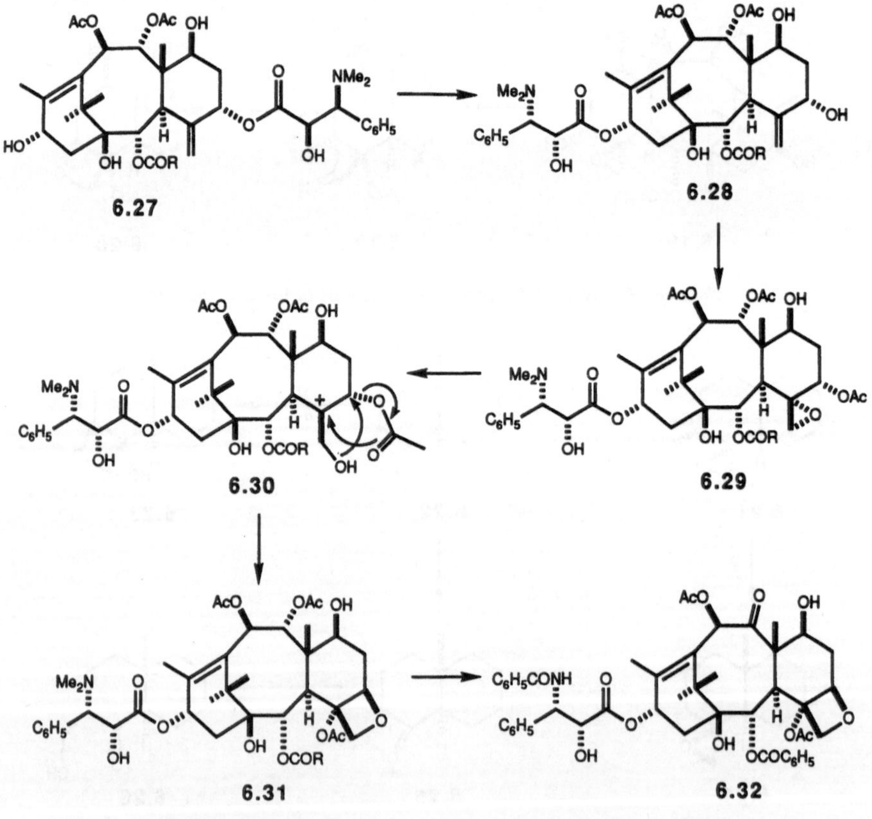

Scheme 6.8. Proposed biogenetic conversion of taxine A to taxol

C-5 and C-13 appear to be widely separated in the conventional representation of the taxane skeleton, they are in fact relatively close in the
3-dimensional structure (Fig. 1) and acyl transfer as indicated is thus
quite possible. Elaboration of the allylic alcohol (6.28) to taxol is
proposed to proceed through the epoxy acetate (6.29), which is suggested
to rearrange to the oxetane (6.31) through the cation (6.30) (Scheme 6.8).

It can thus be seen that several plausible biosynthetic schemes for
both the taxane skeleton and the oxetane ring have been proposed. Of
those schemes that have been tested experimentally on model systems,
few have proven themselves viable and further experimental evidence is
desperately needed to settle the interesting and important questions
raised by these studies.

6.2. Biotransformation of taxoids

The biotransformation of taxol is an important area of study because
a knowledge of its mammalian metabolism is an essential feature of its
clinical pharmacology. In addition, biotransformation of taxol and other
taxanes offers potential pathways to the preparation of chemically
inaccessible metabolites.

Analysis of taxol in human plasma has been carried out to a limited
extent. In two cases (162, 293) no metabolites was detected, but in the
third case (218) a new peak eluting before taxol was detected by HPLC in
the plasma of a cancer patient. Since 7-epitaxol elutes after taxol in their
HPLC system, it was assumed that the peak represents a new metabolite
which is an yet unidentified. Treatment of taxol with cell culture media in
the presence or absence of cells does, however, convert it in part to
7-epi taxol; this conversion is reversible (216).

Metabolism of taxol in rat bile has also been described (184). Only
10% of the injected taxol was recovered in urine, and no urinary
metabolites were detected. On the other hand, over 40% of the injected

6.33

6.34

taxol was found in rat bile, either unchanged or as one of nine metabolites detectable by HPLC. The only metabolite that had lost the side
chain was baccatin-III, indicating that hydrolysis of the C-13 ester side
chain is not a major degradation pathway of taxol. Two of the metabolites were identified as the hydroxylated derivatives (6.33) and (6.34).

No studies of the metabolism of taxol by microbial or other biological systems have appeared so far, but it is probable that several such
studies are ongoing and will yield important information in the future.

7. Bioactivity of Taxol and Other Taxoids

7.1. Toxicity of Taxus Alkaloids

As noted earlier, pharmacological interest in the yew until 1971
centered largely around its toxicity. It has been reported that the yew was
listed in Avicenna's cardiac drugs and was still in use in India in 1970
(284). The toxicity of Taxus extracts is due in large measure to the
presence of taxines and most studies to date have used a mixture of these
alkaloids.

The taxine mixture obtained by WINTERSTEIN (299) was shown to be
toxic, causing convulsions, fall of blood pressure, and stopping of the
heart in diastole. For rabbits the lethal dose was 4–5 mg/kg intravenously and 24 mg/kg by mouth, although a later paper (298) gave a value
of 2 mg/kg as the lethal dose by injection. The taxine mixture prepared
by a different method (28) was considered to be somewhat purer on the
basis of several criteria. It had a toxicity of 2–3 mg/kg in rabbits and
12 mg/kg in mice when injected intravenously and was found to cause
death from asphyxia due to cardiac and respiratory failure (24). A more
recent investigation, but again using a crude mixture of alkaloids
(270, 271), gave median lethal doses of 13 mg/kg for taxine sulfate in the
mouse by subcutaneous injection, 21.9 mg/kg intraperitoneally and
19.7 mg/kg orally. In rats the figure was 20.2 mg/kg for subcutaneous
injection. It has also been shown that taxine acts at the cellular level by
inhibiting both the soidum and calcium current in the cardiac cell
membrane (273). Studies with frog heart preparations showed that taxine
slowed both atrial and ventricular rates, and it was suggested that it acts
as a Ca^{5+} antagonist. A recent study (272) showed that taxine exerted a
cardioprotective effect in rats treated with isoproterenol, a drug which
induces necrosis of the cardiac tissue. This effect may result from the
action of taxine as a calcium antagonist. All this evidence is consistent

with the observation that the fatal symptoms in human subjects who ingest yew leaves are those of cardiac and respiratory failure (*101*).

One study of the pharmacology of the purified taxines A and B has been reported (*3*). Taxine A (**1.3**) has no effect on rats when injected at 5 mg/kg or on guinea pigs when infused at 62 mg/kg/hr. Taxine B (**1.4**), on the other hand, has an LD_{50} of 45 mg/kg in the rat, while its effect on the electrocardiogram of the cat can be detected at 0.53 mg/kg. It thus seems clear that the primary toxic component of the mixture is taxine B, and that this is responsible for the heart damaging activity of the crude taxine.

Other biological activities besides toxicity have also been reported for yew. Thus an aqueous extract of *T. baccata* leaves has anti-ovulatory activity in the rabbit (*33*), and this same extract is also reported to show tranquilizing effects on the central nervous system (*284*). The constituent(s) responsible for these activities have not been identified, but a recent study has shown that several simple taxoids inhibited DNA and protein synthesis in tumor cells (*86*).

7.2. Biological Activity of Taxol and Related Compounds

7.2.1. Antitumor Activity of Taxol

The driving force behind the resurgence of interest in the taxoids in recent years has been the demonstration of taxol's clinical effectiveness in treating human cancer. Since this aspect of taxol has been discussed in several recent reviews (*17, 131, 135, 222, 223, 253, 254*), this section of the present review will simply summarize the key findings in this area.

Taxol was initially isolated because of its strong cytotoxicity and its good activity in the P-388 mouse leukemia assay. However, it was not strongly considered for development as an anticancer agent for several years because of its low aqueous solubility and its difficult supply problem and because its activity appeared to be mainly against the leukemias. This situation changed in 1977, however, when it was shown that taxol had excellent activity against the B16 melanoma and the MX-1 mammary xenograft in nude mice, and taxol was selected at that time for development as an anticancer agent (*253, 254*).

In 1979 HORWITZ reported taxol's unique activity as a tubulin polymerization promoter (*232*). Tubulin is the cellular protein which polymerizes reversibly to form microtubules, which constitute the mitotic spindle apparatus. Taxol binds to microtubules and stabilizes them, thus disrupting the tubulin-microtubule equilibrium and leading to mitotic spindle dysfunction. This effect, which is unique to taxol and related analogs, served to heighten interest in its development as an anticancer agent.

The formulation of taxol was a difficult problem because of its low water solubility. Currently, it is administered in a surfactant formulation with normal saline, using ethanol and Cremophor EL, a polyethoxylated castor oil, as adjuvants (*188, 254*). However, this route of administration is far from ideal since the large doses of taxol required result in the patient receiving relatively large amounts of Cremophor EL. These doses have provoked serious allergic reactions in some cases, both with other drugs (*114*), and with taxol (*148*). It is not known with certainty whether the hypersensitivity reactions in the case of taxol are due to the drug or to the excipient, but their duration and severity can be reduced by premedication with glucocorticoids and antihistamines. Guidelines to prevent or minimize hypersensitivity reactions have been developed (*286*).

Phase I clinical trials of taxol began in the early 1980's, and various protocols were used to evaluate its effects on solid tumors and adult leukemia (*58, 85, 148, 156, 192, 220, 292*). These studies permitted determination of the optimal dose and protocol for Phase II administration, and also showed that taxol gave some partial response in four of 12 patients with melanoma (*292*). The dose-limiting toxicity of taxol is leukopenia, with other toxicities being nausea and vomiting, stomatitis, and the allergic reactions noted above. Recently, it has been noted that cardiac disturbances also accompany taxol treatment in some cases, although here again it is not known whether these are due to taxol itself or to the Cremophor EL surfactant. Cardiac monitoring of patients with known cardiac disease is proposed to alleviate risks from this activity (*224*).

Phase II trials of taxol began in 1985 and are still ongoing, but the results to date have been spectacular in comparison with other anticancer drugs, in spite of the limited trials possible because of the limited supply of drug. The most exciting results have been observed in advanced ovarian epithelial neoplasms, with a response rate of 40% in a group of 40 patients, including one complete response (*177*). Particularly encouraging was the fact that these responses were observed in heavily pretreated patients, and also in patients who were considered to be resistent to cisplatin. Preliminary reports from two other groups confirm the effectiveness of taxol in treating advanced and refractory ovarian cancer (*65, 274*). Very recently, excellent positive responses have also been reported in the treatment of breast cancer with taxol (*105*).

The responses of other cancers to taxol have been less spectacular, but even here there is some encouragement. Thus objective response rates of 12% and 18% have been observed in two separate studies of melanoma, with the latter group including 7% complete response (*64, 155*), and some responses in treatment of non-small-cell lung cancer have been reported. No responses were observed, however, in 18 patients treated for renal cell carcinoma (*63*).

In summary, the data available to date clearly indicate that taxol is an effective drug against ovarian cancer and breast cancer and it seems probable that further studies will detect its activity against other cancers. Taken all in all, it is the most exciting new anticancer drug to be developed in the last twenty years. Recently, data on the antitumor activity of the taxol analogue taxotere (**4.17**) has begun to appear (*16*); this compound shows even better activity than taxol in the B16 melanoma assay in mice. It is thus probable that clinical data with taxotere will also show excellent activity for this compound.

7.2.2. Microtubule Assembly Activity of Taxol

As noted earlier, taxol was shown in 1979 to be a promoter of the assembly of tubulin into microtubules (*232*). This initial key finding has been followed by a flood of papers in which taxol was used as a tool for the isolation and study of microtubules; indeed such papers outnumber those dealing with the chemistry of taxol or of related taxoids.

Since a discussion of taxol's mechanism of action is not the primary focus of this review, and since other reviews have covered the subject in more detail (*17, 112, 171, 254*), this section will be abbreviated. The major findings are that taxol enhances both the rate and yield of microtubule assembly and that the resulting microtubules are resistant to depolymerization by cold, calcium ions, or dilution, which normally destabilize microtubules (*232*). Taxol appears to act as an antimitotic agent by binding to microtubules and then stabilizing them to prevent their normal physiological turnover (*233*). It also eliminates the need for microtubule organizing centers by lowering the critical concentration of tubulin necessary for polymerization, resulting in polymerization of tubulin at many sites in addition to the organizing centers (*45*). Tubulin in its polymer form appears to be the target for taxol, since taxol's biochemical effects on microtubules formed from microtubule protein (tubulin plus microtubule-associated proteins) are also observed with polymer formed from pure tubulin, and taxol can eliminate the requirement for microtubule-associated proteins in microtubule formation (*92, 234*).

The fact that microtubules assembled to steady state and then incubated with taxol become resistant to depolymerization with calcium suggests that there is a binding site for taxol on the microtubule (206) and independent evidence shows that 7-acetyltaxol (which assembles tubulin in a similar fashion to taxol) does not bind to unassembled tubulin (268). Binding saturation on the microtubule occurs at approximate stoichiometry with the tubulin dimer concentration and other drugs that bind to tubulin such as podophyllotoxin and vinblastine do not displace [³H] taxol from microtubules, while unlabeled taxol does. It is clear that the binding site for taxol differs from the site(s) of binding of other antimitotic drugs, since taxol binds only to microtubules while other tubulin active drugs bind only to unpolymerized tubulin. The binding of taxol to assembled microtubules is also non-covalent and reversible (206).

Almost all the evidence available to date suggests that the anticancer activity of taxol is related to its ability to interact with polymerized tubulin. Thus MANFREDI and HORWITZ could write in 1984: "The effects of taxol have been studied in a number of different systems, and in all cases examined the effects of taxol appear to be related specifically to the tubulin-microtubule system" (171). These authors also suggested that the ability of plant alkaloids such as taxol and colchicine to interact with tubulin indicates the presence of endogenous molecules in animal cells with similar activites. A further indication of the relationship between tubulin-assembly activity and anticancer activity is that cytotoxicity and tubulin-assembly activity broadly parallel each other in a range of taxol analogs that have been investigated; the few examples of compounds that do not show parallel activity can be explained in reasonable ways (131).

The connection between taxol's tubulin-assembly activity and its anticancer activity has been suggested to be that taxol prevents the cell from depolymerizing its microtubule cytoskeleton (233). This activity is, however, not necessarily related to the inhibition of mitotic spindle formation; it has been proposed that taxol alters calcium regulation in the cell by interacting with membrane-bound tubulins of cytoplasmic organelles (91). Since calcium is involved in the microtubule-tubulin equilibrium, this alteration of calcium regulation could also involve disruption of the mitotic spindle function.

It is also possible that taxol's action as an anticancer agent is due to some other effect of its action on tubulin. The spectrum of antitumor activity of taxol is very different from that of other tubulin-active drugs such as vinblastine, vincristine, colchicine, podophyllotoxin, and maytansine. These compounds act primarily as antileukemic agents, whereas taxol acts mainly on solid tumors. It is possible that the effects of taxol on calcium ion fluxes may be involved in cellular signalling mechanisms, or that its primary action might be on tubulin/microtubule isotypes that are different from those involved in mitosis.

In summary, taxol's unique ability to promote the assembly of tubulin by binding to microtubules appears to be at the core of its important anticancer activity, but it is not at present clear how this ability translates into the actual activity observed.

7.2.3. Structure-activity Relationships of Taxol Analogs

A large number of taxol analogs have been prepared in attempts to understand structure-activity relationships in this area. Since these relationships have recently been reviewed (131), this section will be abbreviated.

The major conclusions to emerge from the available data are that the C-13 ester side chain is essential for activity, but that the actual chemical structure of the side chain can be varied to some extent while retaining activity. Thus some semi-synthetic taxol analogs with modified side chains show activites comparable to that of taxol and the semisynthetic analog taxotere is even slightly more active than taxol in some assays (88, 217, 263). Fewer analogues are available for ring-modified taxols, but it has been shown that the oxetane ring is essential for activity, and a rearranged taxol analog which retains the oxetane ring shows tubulin assembly activity (but not cytotoxicity) comparable to that of taxol (230).

Addendum

Since the submission of the manuscript in April 1992; a number of additional papers on the taxane diterpenoids have appeared whose key features are summarized below.

3. The Families of Taxane Diterpenoids

An ethanolic extract of the bark of *Taxus yunnanensis* (*S11, 12*) afforded eight taxoids, 7 of which were isolated previously and were identified as taxinine E, taxinine J, 1β-acetoxy-5α-deacetylbaccatin I, baccatin III, taxol, cephalomannine, and 7-(β-xylosyl)-10-deacetyltaxol. The eighth taxoid is new and was named yunnanxane (**3.S1**).

Four other taxoids have been recently isolated from *Taxus* chinensis (*S28*). Two of them are the known 5α-cinnamoyloxy-2α,13α-dihydroxy-9α,10β-diacetoxy-4 (20), 11-taxadiene and 5α-cinnamoyloxy-10β-hydroxy-2α,9α,13α-triacetoxy-4 (20), 11-taxadiene. The other two, (**3.S2, 3.S3**) are shown below.

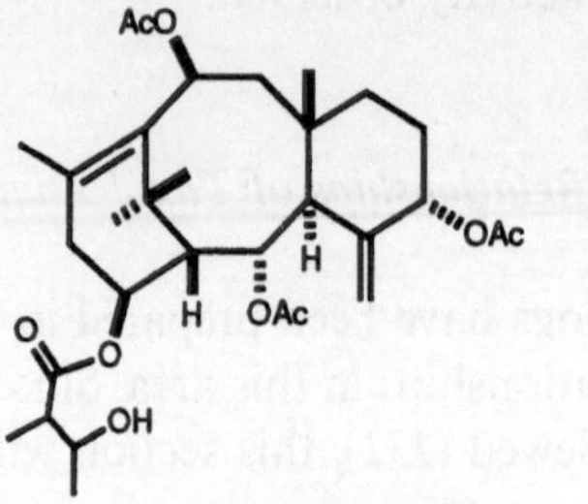

$C_{31}H_{46}O_9$; MW=562
mp °C = 165-167
$[\alpha]_D$ = +41.6° (MeOH)

Yunnanxane (**3.S1**)

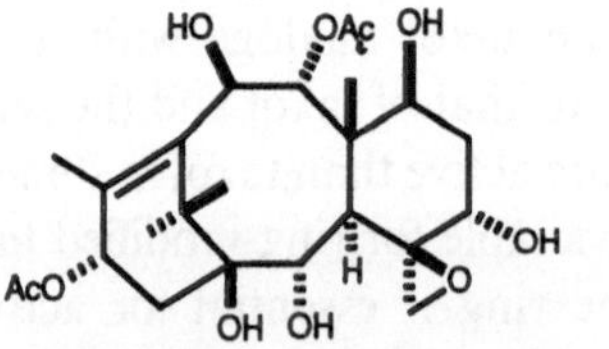

$C_{24}H_{36}O_{10}$; MW=484
mp °C = 255-257
$[\alpha]_D$ = +228.9° (CHCl₃)

1β-Hydroxy-2α,5α,7β,10β-
tetradeacetylbaccatin I (**3.S2**)

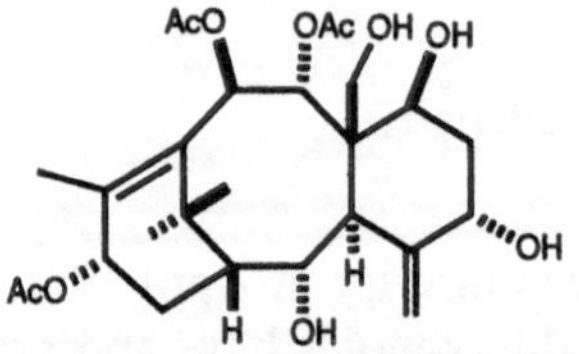

$C_{26}H_{38}O_{10}$; MW=510
mp °C = 223-225
$[\alpha]_D$ = -20.4° (CHCl₃)

2α,5α,Dihydroxy-19-hydroxymethyl-
9α,10β,13α-triacetoxy-4(20),11-taxadiene (**3.S3**)

Several papers have recently appeared from APPENDINO and his collaborators, who studied side cuts from the commercial production of 10-deacetylbaccatin III. Thus the needles of *T. wallichiana* Zucc. gave a new taxane whose structure was established as 14β-hydroxy-10-deacetylbaccatin III on the basis of chemical reactions, spectroscopic data, and X-ray analysis (*S1*), whereas *T. baccata* needles afforded a taxane diester characterized by an additional bond between C-3 and C-11. The structure was elucidated by spectral methods and further confirmed by synthesis from a known taxane derivative. A general procedure for the photocyclization of taxicins is also reported (*S3*). A winter collection of yew needles from *T. baccata* gave two new pairs of taxicin I and taxicin II derivatives (*S2*) while the seeds of *T. baccata* gave two basic taxanes related to taxine B and an oxygenated derivative of brevifoliol. Conformational flexibility of the tricyclic taxane skeleton was detected in the latter compound (*S4*).

A Canadian group has isolated three new taxanes from the needles and stems of *T. canadensis* together with 5 known taxanes (*S53*). The new taxanes were identified as 7,9-deacetylbaccatin VI, 1-acetyl-10-deacetylbaccatin III, and 1β-hydroxy-7,9-deacetylbaccatin I.

Information on some Taiwanese and Chinese work on new taxoids has recently become available. A reinvestigation of the constituents of the heartwood of Taiwan yew yielded 9α,10β-diacetoxy-5α,13α-dihydroxy-4(20),11-taxadiene, and 5α,7β,10β-triacetoxy-2α-(α-methylbutyryloxy)-4(20)11-taxadiene in addition to other compounds (*S13*). A later paper by the same authors added 1-dehydroxybaccatin IV to the taxoids isolated (*S14*). The Zhang group continues to isolate new taxoids and has reported four new taxoids from *T. chinensis*. These are 9α,10β-diacetoxy-2α-benzoyloxy-1β,5α,13α-trihydroxy-4(20)11-taxadiene, 9α,10β-diacetoxy-2α-benzoyloxy-5α-cinnamoyloxy-1β,13α-dihydroxy-4(20), 11-taxadiene, 9α,10β,13α-triacetoxy-2α-benzoyloxy-1β, 5α-dihydroxy-4(20), 11-taxadiene, and 1β,2α,5α,9α,10β,13α-hexahydroxy-4(20),11-taxadiene (*S55*).

Two new taxoids were isolated from *T. mairei* and identified as 5α,7β,9α,10β,13α-pentaacetoxy-2α-benzoyloxy-4α,20-dihydroxytaxa-11(12)-ene and the isomeric 7β,9α,10β,13α,20-pentaacetoxy-2α-benzoyloxy-4α,5α-dihydroxytaxa-11(12)-ene (*S31*). The reduced baccatin III analogue 9-dihydro-13-acetylbaccatin III was isolated from *T. canadensis* (*S23*).

4. The Chemistry of the Taxoids

Several papers on the chromatography of taxol and related taxoids have appeared. Taxol can be isolated from prepurified extracts of the stem bark of *T. baccata* by high-speed countercurrent chromatography; a mixture of taxol and cephalomannine is obtained and this mixture can be partially separated (*S47*). Taxol and cephalomannine can be completely separated by reversed-phase HPLC on a microcolumn packed with octadecyl silica (*S24*). Analysis of *T. chinensis* extracts for taxol has also been carried out by HPLC on a silica gel column using CH_2Cl_2–MeOH as the mobile phase (*S52*). Taxol has also been determined in needle and bark samples of *T. cuspidata* by a combination of HPLC with thermo-spray mass spectrometry (*S5*). The method provides good sensitivity and is linear over the range 1–1000 ng of taxol per injection. The application of HPLC to the analysis of bark and foliage samples of *T. brevifolia* grown under either shady or sunny conditions confirmed that taxol and cephalomannine concentration was highest in the bark and showed that shaded trees yielded significantly more of these taxoids than sun-exposed trees (*S29*).

An additional NMR study of taxol has appeared (*S6*). This study used nuclear Overhauser effect spectroscopy and molecular mechanics calculations to arrive at a structure for taxol which was very similar to the published crystal structure of taxotere except that the two ends of the molecule are slightly closer together than in the X-ray structure.

Treatment of 7,13-diacetyl baccatin III with tributyltin methoxide in the presence of lithium chloride led to the formation of the isobaccatin III derivative **4.S1** (*S18*). A similar rearrangement has also been observed with the fully deacylated baccatin III derivative **4.40**; treatment with either acid or base yields the isobaccatin derivative **4.S2** (*S44*). Various rearrangement reactions have also been observed with 10-deacetylbaccatin III. These include reactions that yield products similar to isobaccatin **4.S2** and to the A-nortaxol **4.90** as well as other products (*S48*).

4.S1   **4.S2**

5. Approaches to the Synthesis of Taxane Diterpenoids

In the area of the total synthesis of taxoids, a number of publications have appeared. FÉTIZON's group has extended the scope of their investigation using [2 + 2] photocycloaddition-fragmentation chemistry to construct **AB** intermediates (*S7*). This extension focused on the photocycloaddition of vinyl acetate to a 1,3-dione enol ether. The result of this work, was the preparation of the bicyclo[5.2.1]dodecane derivative (**5.S1**).

PAQUETTE's group has continued their investigation on the application of anionic oxy-Cope rearrangements to the enantiospecific synthesis

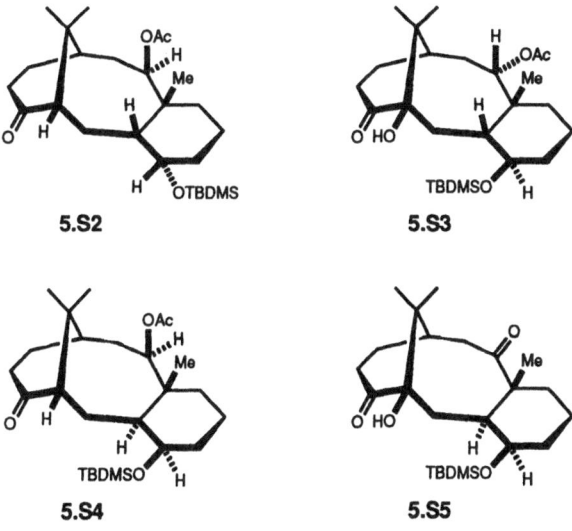

5.S1

of tricyclo[9.3.1.0^{3,8}]-pentadecanes suitable for the eventual preparation of natural taxoids. One report (*S40*) describes the application of α-ketol and Wagner-Meerwein rearrangements to give direct access to functionalized cis-tricyclo[9.3.1.0^{3,8}]pentadecanes (**5.S2**) and (**5.S3**). Taxoid intermediates (**5.S2**) and (**5.S3**) were prepared in 7 and 8 steps, respectively, from 2-oxo-7,7-dimethyl-1-vinylbicyclo[2.2.1]heptane.

An accompanying report (*S39*) details the results of an investigation designed to explore the consequences of setting the requisite *trans* relationship in intermediates prior to the 1,2-shift that will deliver taxoid precursors like (**5.S2**) and (**5.S3**). This study led to in the preparation of (**5.S4**), a potential taxusin precursor, in twelve steps, and (**5.S5**), a potential taxol precursor, in ten steps.

An additional publication (*S38*) describes the preparation of enantiomerically pure triketone (**5.S6**) from (*R*)-2-oxo-7,7-dimethyl-1-vinylbicyclo[2.2.1]heptane in eight steps and 40% overall yield. This report also describes an unusual redox cascade that occurred with a taxoid derivative.

5.S2

5.S3

5.S4

5.S5

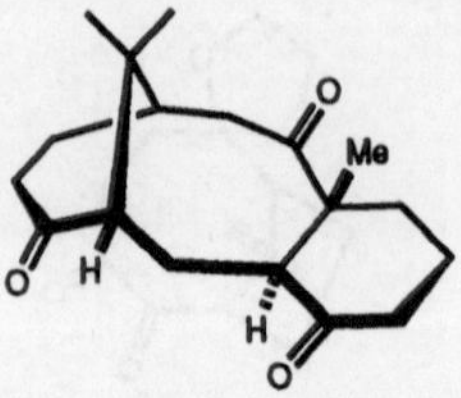

5.S6

The DANISHEFSKY group has recently introduced their approach to the synthesis of taxoids. Their initial publication (*S32*) outlines the preparation of functionalized intermediates containing the CD substructure of taxol. This report describes the preparation of CD precursors (**5.S7**) and (**5.S8**) from the Wieland-Miescher ketone.

A second report (*S42*) describes the construction of B-*seco* taxoid derivatives (**5.S9**) and (**5.S10**). Their approach to these taxoid precursors relied on the preparation of A-ring intermediates appropriately functionalized so that they might function as dieneophiles to provide rapid access to the *seco*-B-taxoid analogues.

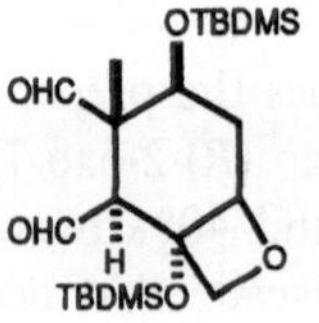

5.S7

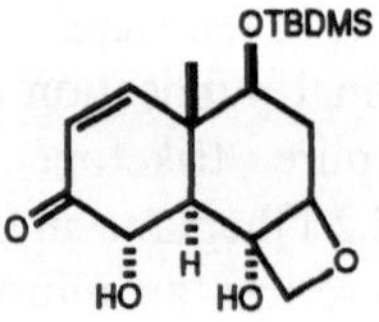

5.S8

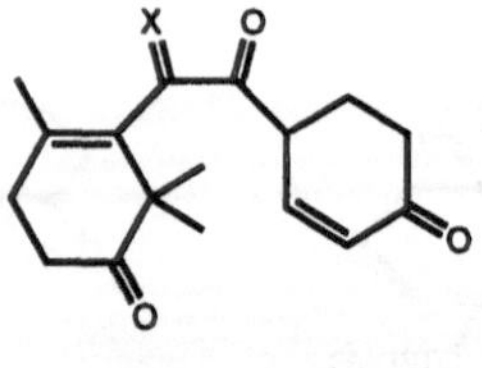

5.S9 X = H,H
5.S10 X = O

WENDER's group has reported (*S49*) an approach to the taxoids that is strategically different from the one they adopted originally. This new approach, described as "The Pinene Path", is outlined in Scheme 5.S1. Key intermediate (**5.S12**) was prepared from pinene (**5.S11**) in four steps. Application of a fragmentation similar to that developed by Holton (106,109) provided tricycle (**5.S13**). A three step sequence was then used to convert (**5.S13**) into taxoid intermediate (**5.S14**). The advantage of this

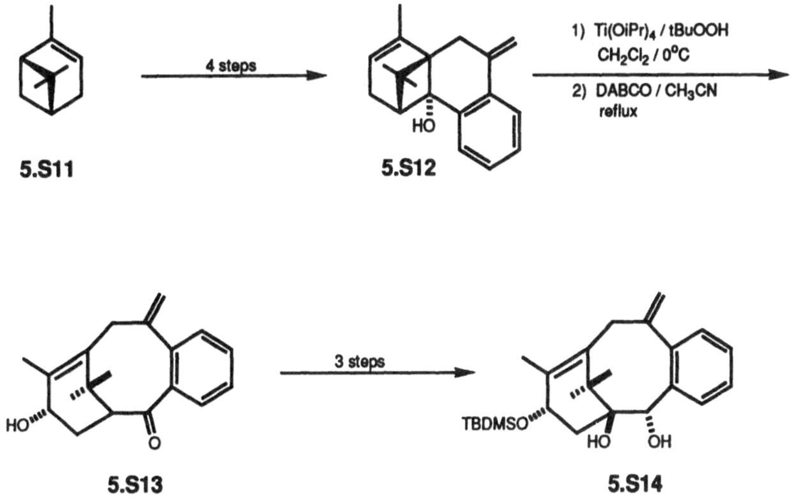

Scheme 5.S1. Wender's new approach to the synthesis of taxoids

new route is that it allows the construction of analogues and potential precursors of taxol in the correct enantiomeric form using readily available and inexpensive starting material. This group has also reported an approach to the C,D ring system of taxol by photolysis of α-methoxy ketones (*S50*).

BLECHERT *et al.* have expanded their work to include the stereoselective synthesis of a tricyclic taxoid system with three oxygen centers in ring B (*S9*). The same strategy was used to prepare a variety of functionalized A, B-ring systems.

SHEA and coworkers have reported the atroposelective synthesis of tricyclo[9.3.1.0^{3,8}]pentadecane ring systems (*S27*). This work also describes the use of substrate conformation to stereoselectively elaborate functional groups. Also disclosed were the results of an investigation designed to functionalize the C-1 position (*S45*).

WINKLER's group have expanded their application of the intramolecular dioxenone photocycloaddition to the synthesis of C-13 oxygenated taxoid analogs (*S51*). This work includes attachment of the taxol side chain to the tricyclic analogs that were prepared, but unfortunately bioactivity data was not available.

PATTENDEN and HITCHCOCK have recently disclosed a new approach to the taxoid skeleton (*S25*). This route involves a tandem radical macrocyclization-radical transannulation sequence.

The most recent report of KUWAJIMA and coworkers gives the full details of earlier work that resulted in the preparation of C-aromatic taxoid models (*S20*). FUNK *et al.* have also continued their investigation of the Claisen rearrangement to prepare BC intermediates (*S19*).

The first two papers on a convergent strategy for taxol have appeared from the NICOLAOU group (*S35*, *S36*), and report the synthesis of fully functionalized rings A and rings C/D of taxol.

Two groups have reported syntheses of bicyclic portions of the taxane skeleton. FETIZON and coworkers prepared a B/C fragment by a photo-chemical approach (*S8*), and GHOSH and coworkers prepared an A/B fragment (*S43*). The taxane skeleton was also used as an example in a paper describing a computer-based approach to finding key steps in the synthesis of complex structures. The program described 15 routes involving an oxy-Cope reaction and 11 routes using a de Mayo reaction (*S34*).

There continues to be great interest in the synthesis of taxol and taxol analogues from baccatin III, fueled both by the hope of obtaining analogues with better activity and by the desire to develop improved synthetic methods. The great example in much of this work has of course been taxotere, and a short account of its discovery has appeared (*S41*).

Four syntheses of protected taxol side chains have appeared. JACOB-SEN and DENG have reported an enantioselective four step synthesis of the side chain starting from commercially available ethyl phenylpropiolate.; the key step was the enantioselective epoxidation of *cis*-ethyl cinnamate, and the overall yield of this route was 25% (*S16*). FARINA *et al.* have disclosed a chiral β-lactam approach to the side chain. This approach was based on the Staudinger reaction, and uses an L-threonine ester as the inducer of chirality (*S17*). OJIMA's group has also reported a chiral β-lactam route to the taxol side chain. Their route is based on a chiral ester enolate-imine condensation, and they included an improved proced-ure for coupling the β-lactam with a protected baccatin III (*S37*). COMMERCON and coworkers have described an improved synthesis of the side chain that makes use of EVANS' chiral oxazolidinone methodology to induce chirality, and goes on to protect the final product as an oxazoli-dine. Coupling to protected baccatin III proceeded in high yield, as did the deprotection and reacylation steps necessary to make taxol or taxotere (*S15*).

The synthesis of two *p*-chlorophenyl analogues of taxol has been reported: the products which had a *p*-chlorine on the N-benzoyl or the 3'-phenyl ring of taxol, had activity comparable to that of taxol (*S21*, *S22*). A photoreactive taxol side chain has also been prepared (*S10*).

6. Biosynthesis and Biotransformation of Taxoids

An abstract of a biosynthetic study of taxol has appeared (*S30*). Feeding of $[7\text{-}^{14}C]$benzoic acid to cut stems of *T. brevifolia* resulted in significant incorporation of $[^{14}C]$ into taxol, and several cell lines of *T. brevifolia* have been established in suspension culture. A recent paper has confirmed that the taxane ring system derives from mevalonate and has shown that the side chain derives from phenylalanine. The acetate groups, not unexpectedly, derive from acetate (*S54*).

7. Bioactivity of Taxol and Other Taxoids

A review on the anti-cancer activity of taxol has appeared (*S46*). In a brief review of it, mechanism of action has also been published.

Acknowledgements

The authors thank Professors K. NAKANISHI (Columbia), B. SNIDER (Brandeis), I. KUWA-JIMA (Tokyo Institute of Technology), L.A. PAQUETTE (Ohio State), and P. WENDER (Stanford), who sent preprints or reprints of their work on taxane diterpenoids. The ^{1}H-NMR spectrum of taxol (Fig. 2) was kindly provided by Drs. G.N. CHMURNY and J. BEUTLER of the NCI-Frederick Cancer Research and Development Center, and Professor G. PATTENDEN (Nottingham) provided information about the correct stereochemistry of verticillol. The authors thank Dr. M. SUFFNESS and Dr. K. SNADER of the National Cancer Institute who read a draft of the manuscript and made helpful suggestions.

References

1. ANDRIAMIALISOA, R.Z., M. FETIZON, I. HANNA, C. PASCARD, and T. PRANGE: Synthetic Studies on Taxane Diterpenes. X-Ray Structure of a Key Intermediate. Tetrahedron **40**, 4285 (1984).

2. BALZA, F., S. TACHIBANA, H. BARRIOS, and G.H.N. TOWERS: Brevifoliol, A Taxane from *Taxus brevifolia*. Phytochemistry **30**, 1613 (1991).

3. BAUEREIS, R., and W. STEIERT: Pharmakologische Eigenschaften von Taxin A und B. Arzneimittel Forsch. **9**, 77 (1959).

4. BAXTER, J. N., B. LYTHGOE, B. SCALES, and S. TRIPPETT: Taxine-I, The Major Alkaloid of the Yew, *Taxus baccata* L. Proc. Chem. Soc. 9 (1958).

5. BAXTER, J.N., B. LYTHGOE, B. SCALES, R.M. SCROWSTON, and S. TRIPPETT: Taxine. Part I. Isolation Studies and the Functional Groups of O-Cinnamoyltaxicin-I. J. Chem. Soc. 2964 (1962).

6. BEGLEY, M.J., E.A. FRECKNALL, and G. PATTENDEN: X-ray Structure of O-Cinnam-oyltaxicin-I Triacetate, $C_{35}H_{42}O_{10}$, from the Yew, *Taxus baccata*. Acta Cryst. **C40**, 1745 (1984).

7. BEGLEY, M.J., C.B. JACKSON, and G. PATTENDEN: Investigation of Transannular Cyclisations of Verticillanes to the Taxane Ring System. Tetrahedron Letters **26**, 3397 (1985).

8. BEGLEY, M.J., C.B. JACKSON, and G. PATTENDEN: Total Synthesis of Verticillene. A Biomimetic Approach to the Taxane Family of Alkaloids. Tetrahedron **46**, 4907 (1990).

9. BENCHIKH-LE-HOCINE, M., D. DO KHAC, H. CERVANTES, and M. FETIZON: Model Studies in Taxane Diterpene Synthesis. Part II. New Journal of Chem. **10**, 715 (1986).

10. BENCHIKH-LE-HOCINE, M., D. DO KHAC, M. FETIZON, I. HANNA, and R. ZEGHDOUDI: A Photochemical Approach to the Bicyclo[5.3.1] Undecane Ring System of Taxane Diterpenes. Syn. Commun. **17**, 913 (1987).

11. BENCHIKH-LE-HOCINE, M., D. DO KHAC, M. FETIZON, F. GUIR, Y. GUO, and T. PRANGE: Une Nouvelle Voie d'Acces Aux Derives du Taxane. Tetrahedron Letters **33**, 1443 (1992).

12. BERKOWITZ, W.F., and A.S. AMARASEKARA: An Approach to the D-Ring of Baccatin III (Taxol, Cephalomannine). Tetrahedron Letters **26**, 3663 (1985).

13. BERKOWITZ, W.F., A.S. AMARASEKARA, and J.J. PERUMATTAM: A Photochemical Approach to the Taxanes. J. Org. Chem. **52**, 1119 (1987).

14. BERKOWITZ, W.F., J. PERUMATTAM, and A. AMARASEKARA: A Photochemical Approach to the Taxanes. Tetrahedron Letters **26**, 3665 (1985).

15. BEUTLER, J.A., G.M. CHMURNY, S.A. LOOK, and K.M. WITHERUP: Taxinine M, A New Tetracyclic Taxane from *Taxus brevifolia*. J. Nat. Prod. **54**, 893 (1991).

16. BISSERY., M.-C., D. GUÉNARD, F. GUÉRITTE-VOEGELEIN, and F. LAVELLE: Experimental Antitumor Activity of Taxotere (RP 56976, NSC 628503), a Taxol Analogue. Cancer Res. **51**, 4845 (1991).

17. BLECHERT, S., and D. GUÉNARD: Taxus Alkaloids. In: The Alkaloids, Vol. 39, Chemistry and Pharmacology. Ed. A. BROSSI, p. 195. New York: Academic Press. 1990.

18. BLECHERT, S., and A. KLEINE-KLAUSING: Synthesis of a Biologically Active Taxol Analogue. Angew. Chem. Int. Ed. **30**, 412 (1991).

19. BLUME, E.: Investigators Seek to Increase Taxol Supply [News]. J. Natl. Cancer. Inst. **81**, 1122 (1989).

20. BONNERT, R. V., and P.R. JENKINS: A New Synthesis of Substituted Dienes and its Application to an Alkylated Taxane Model System. J. Chem. Soc. Chem. Commun. 1540 (1987).

21. BONNERT, R.V., and P.R. JENKINS: A Synthesis of an Alkylated Taxane Model System. J. Chem. Soc. Perkin Trans. I 413 (1989).

22. BROWN, P.A., and P.R. JENKINS: Synthesis of the Taxane Ring System Using an Intramolecular Diels-Alder Reaction of a 2-Substituted Diene. J. Chem. Soc. Perkin Trans. I 1303 (1986).

23. BROWN, P.A., P.R. JENKINS, J. FAWCETT, and D.R. RUSSELL: A Stereocontrolled Route to the Tricyclo[9.3.1.0$^{3, 8}$]pentadecane Ring System of Taxane. J. Chem. Soc. Chem. Commun. 253 (1984).

24. BRYAN-BROWN, T.: Pharmacological Action of Taxine. Quart. J. Pharm. Pharmacol. **5**, 205 (1932).

25. BÜCCI, G., W.D. MACLEOD, Jr., and J. PADILLA, O.: Terpenes. XIX. Synthesis of Patchouli Alcohol. J. Am. Chem. Soc. **86**, 4438 (1964).

26. BURKE, B.A., W.R. CHAN, V.A. HONKAN, J.F. BLOUNT, and P.S. MANCHAND: The Structure of Cleomeolide, an Unusual Bicyclic Diterpene from *Cleome Viscosa* L. (Capparaceae). Tetrahedron **36**, 3489 (1980).

27. CAESAR, J.: The Battle for Gaul, Book 6, Section 31. Translated by WISEMAN, A. and P. WISEMAN, p. 126. London: Chatto and Windus. 1980.

28. CALLOW, R.K., J.M. GULLAND, and C.J. VIRDEN: Physiologically Active Constituents of the Yew, *Taxus baccata*. Part 1. Taxine. J. Chem. Soc. 2138 (1931).

29. CARDELLINA II, J.H.: HPLC Separation of Taxol and Cephalomannine. J. Liq. Chromatogr. **14**, 659 (1991).

30. CASTELLANO, E.E. and O.J.R. HODDER: The Crystal and Molecular Structure of the Diterpenoid Baccatin V, a Naturally Occurring Oxetan with a Taxane Skeleton. Acta Cryst. **B29**, 2566 (1973).

31. CERVANTES, H., D. DO KHAC, M. FETIZON, F. GUIR, J. BELOEIL, J. LALLEMAND, and T. PRANG: Model Studies in the Taxane Diterpene Series – Part I. Tetrahedron **42**, 3491 (1986).

32. CHAN, W.R., T.G. HALSALL, G. M. HORNBY, A.W. OXFORD, W. SABEL, K. BJAMER, G. FERGUSON, and J.M. ROBERTSON: Taxa-4(16),11-diene-5α,9α,10β,13α-tetraol, a New Taxane Derivative from the Heartwood of Yew (*T. baccata* L.): X-Ray Analysis of a p-Bromobenzoate Derivative. Chem. Commun. 923 (1966).

33. CHAUDHURY, R.R., S.K. SAKESENA, and S.K. GARG: Preliminary Observations in the Rabbit on the Anti-ovulatory Activity Present in *Taxus baccata* Linn. Leaves. J. Reprod. Fertil. **22**, 151 (1970).

34. CHAUVIERE, G., D. GUÉNARD, C. PASCARD, F. PICOT, P. POTIER, and T. PRANGE: Taxagifine: New Taxane Derivative from *Taxus baccata* L. (Taxaceae). J. Chem. Soc., Chem. Commun. 495 (1982).

35. CHAUVIERE, G., D. GUÉNARD, F. PICOT, V. SENILH, and P. POTIER: Analyse Structurale et Etude Biochimique de Produits Isoles de l'If: *Taxus baccata* L.'(Taxacées). C. R. Acad. Sc. Paris, Serie II, **293**, 501 (1981).

36. CHEN, W.M.: Chemical Constituents and Biological Activity of Plants of *Taxus* genus. Acta Pharm. Sin. (Yaoxue Xuebao) **25**, 227 (1990).

37. CHENU, J., M. TAKOUDJU, M. WRIGHT, V. SENILH, and D. GUÉNARD: Synthesis of a [³H]-Labelled Derivative of the Microtubular Poison Taxol. J. Labelled Comp. Radiopharm. **24**, 1245 (1987).

38. CHIANG, H.C.: The Constituents of *Taxus chinensis* Rehd. Shi Ta Hsueh Pao **20**, 147 (1975).

39. CHIANG, H.C., M.C. WOODS, Y. NAKADAIRA, and K. NAKANISHI: The Structures of Four New Taxinine Congeners, and a Photochemical Transannular Reaction. Chem. Commun. 1201 (1967).

40. CHMURNY, G.N., B.D. HILTON, S. BROBST, S.A. LOOK, K.M. WITHERUP, and J.A. BEUTLER: ¹H and ¹³C-NMR Assignments for Taxol, 7-*epi*-Taxol, and Cephalomannine. J. Nat. Prod. **55**, 414 (1992).

41. CLARK, G.R., J. LIN, and M. NIKAIDO: Aldol Reactions of a Cyclobutanone Enolate. Tetrahedron Letters **25**, 2645 (1984).

42. COLIN, M., D. GUÉNARD, F. GUÉRITTE-VOEGELEIN, and P. POTIER: Procédé de Préparation du Taxol et du Désacétyl-10 Taxol. Eur. Patent 0,253,739. January 20, 1988.

43. DAUBEN, W.G., J.P. HUBBELL, P. OBERHANSLI, and W.E. THIESSEN: Acid-Catalyzed Cyclization of Cembrene and Isocembrol. J. Org. Chem. **44**, 669 (1979).

44. DEAMICIS, C.V.: Insertion Reactions of Oxacarbenes Generated Photochemically from Cyclobutanones. Ph.D. Dissertation, Stanford University, 1988.

45. De Brabander, M., G. Geuens, R. Nuydens, R. Willebrords, and J. de Mey: Taxol Induces the Assembly of Free Microtubules in Living Cells and Blocks the Organizing Capacity of the Centrosomes and Kinetochores. Proc. Natl. Acad. Sci. USA **78**, 5608 (1981).

46. Della Casa De Marcano, D.P., and T.G. Halsall: The Isolation of Seven New Taxane Derivatives from the Heartwood of Yew (*Taxus baccata* L.). Chem. Commun. 1282 (1969).

47. Della Casa De Marcano, D.P., T.G. Halsall, E. Castellano, and O.J.R. Hodder: Crystallographic Structure Determination of the Diterpenoid Baccatin-V, a Naturally Occurring Oxetan with a Taxane Skeleton. J. Chem. Soc. D (Chem. Commun.) 1382 (1970).

48. Della Casa De Marcano, D.P., and T.G. Halsall: The Structure of the Diterpenoid Baccatin-I, the 4β,20-Epoxide of 2α,5α,7β,9α,10β,13α-Hexa-acetoxytaxa-4(20),11-Diene. J. Chem. Soc. D (Chem. Commun.) 1381 (1970).

49. Della Casa De Marcano, D.P., and T.G. Halsall: Structures of some Taxane Diterpenoids, Baccatins-III, -IV, -VI, and -VII and 1-Dehydroxybaccatin-IV, Possessing an Oxetan Ring. J. Chem. Soc., Chem. Commun. 365 (1975).

50. Della Casa De Marcano, D.P., T.G. Halsall, and G.M. Hornby: The Structure of Baccatin-III, a Partially Esterified Octahydroxy-monoketo-taxane Derivative Lacking a Double Bond at C-4. J. Chem. Soc. D (Chem. Commun.) 216 (1970).

51. Della Casa De Marcano, D.P., T.G. Halsall, A.I. Scott, and A.D. Wrixon: Application of the Olefin Octant Rule to some Taxane Derivatives: Assignment of Absolute Configuration using the Cotton Effect. J. Chem. Soc. D (Chem. Commun.) 582 (1970).

52. De Marcano, D., B. Mendez, J. De Mendez, J. Monasteriios, A.C. Rojas, and T.G. Halsall: Carbon-13 NMR Spectra of $\Delta^{4(20)11}$-Taxadiene Derivatives. Org. Magn. Reson. **21**, 524 (1983).

53. Denis, J.-N., A. Correa, and A.E. Greene: An Improved Synthesis of the Taxol Side Chain and of RP56976. J. Org. Chem. **55**, 1957, 5538 (1990).

54. Denis, J.-N., A.Correa, and A.E.Greene: Direct, Highly Efficient Synthesis from (S)-(+)-Phenylglycine of the Taxol and Taxotere Side Chains. J. Org. Chem. **56**, 6939 (1991).

55. Denis, J.-N., A.E. Greene, D. Guénard, F. Guéritte-Voegelein, L. Mangatal, and P. Potier: Highly Efficient, Practical Approach to Natural Taxol. J. Am. Chem. Soc. **110**, 5917 (1988).

56. Denis, J.-N., A.E. Greene, A. A. Serra, and M.-J. Luche: An Efficient, Enantioselective Synthesis of the Taxol Side Chain. J. Org. Chem. **51**, 46 (1986).

57. Deutsch, H.M., J.A. Glinski, M. Hernandez, R.D. Haugwitz, V.L. Narayanan, M. Suffness, and L.H. Zalkow: Synthesis of Congeners and Prodrugs. 3. Water-Soluble Prodrugs of Taxol with Potent Antitumor Activity. J. Med. Chem. **32**, 788 (1989).

58. Donehower, R.C., E.K. Rowinsky, L.B. Grochow, S.M. Longnecker, and D.S. Ettinger: Phase I Trial of Taxol in Patients with Advanced Cancer. Cancer Treat. Rep. **71**, 1171 (1987).

59. Dorn, H.C., C. Tsiao, K.H. Tsai, G. Samaranayake, and D.G.I. Kingston: Interfacial Studies; Taxol/Silica Phase Immobilized Nitroxide ^{1}H DNM Intermolecular Transfer Enhancements. J. Magn. Res., submitted (1992).

60. Duchesne, J.P., and M. Mulhauser: Preparation of (2R, 3R)-β-phenylglycidic acid as an Intermediate for Taxol. PCT. Int. Appl. W091 13,066, Sep. 5, 1991 (Chem. Abstr. **115**, 279790u (1991)).

61. DUKES, M., D.H. EYRE, J.W. HARRISON, and B. LYTHGOE: The Stereochemistry of Taxicin-I and II. Tetrahedron Letters 4765 (1965).

62. DUKES, M., D.H. EYRE, J.W. HARRISON, R.M. SCROWSTON, and B. LYTHGOE: Taxine. Part V. The Structure of Taxicin II. J. Chem. Soc. C. 448 (1967).

63. EINZIG, A.I., E. GOROWSKI, J. SASLOFF, and P.H. WIERNIK: Phase II Trial of Taxol in Patients (Pts) with Renal Cell Carcinoma. Proc. Am. Assoc. Cancer Res. **29**, 222 (1988).

64. EINZIG, A.I., D.L. TRUMP, J. SASLOFF, E. GOROWSKI, J. DUTCHER, and P.H. WIERNIK: Phase II Pilot Study of Taxol in Patients (Pts) with Malignant Melanoma (mm). Proc. Am. Soc. Clin. Oncol. **7**, 249 (1988).

65. EINZIG, A.I., P.H. WIERNIK, J. SASLOFF, S. GARL, C. RUNOWICZ, K.A. O'HANLAN, G. GOLDBERG: Phase II Study of Taxol (T) in Patients (Pts) with Advanced Ovarian Cancer. Proc. Am. Assoc. Cancer Res. **31**, 187 (1990).

66. ELMORE, S.W., K.D. COMBRINK, and L.A. PAQUETTE: A Convenient Means for Controlling the Oxidation Level of Bridgehead Carbon C-1 in Functionalized Tricyclo[9.3.1.0^{3,8}]pentadecanes. Tetrahedron Letters **32**, 6679 (1991).

67. ERDTMAN, H., and K. TSUNO: *Taxus* Heartwood Constituents. Phytochem. **8**, 931 (1969).

68. ETTOUATI, L., A. AHOND, O. CONVERT, D. LAURENT, C. POUPAT, and P. POTIER: Plantes de Nouvelle-Calédonie. 114. Taxanes isoles des feuilles d'*Austrotaxus spicata* Compton (Taxacées). Bull. Soc. Chim. France 749 (1988).

69. ETTOUATI, L., A. AHOND, O. CONVERT, C. POUPAT, and P. POTIER: Plantes de Nouvelle-Caledonie. 124. Taxanes isoles des ecores de tronc d'*Austrotaxus spicata* Compton (Taxacées). Bull. Soc. Chim. France 687 (1989).

70. ETTOUATI, L., A. AHOND, C. POUPAT, and P. POTIER: Revision Structurale de la Taxine B, Alcaloide Majoritaire des Feuilles de l'If d'Europe, *Taxus baccata*. J. Nat. Prod. **54**, 1455 (1991).

71. ETTOUATI, L., A. AHOND, C. POUPAT, and P. POTIER: Première Hémisynthèsis d'un Composé de Type Taxane Porteur d'un Groupement Oxétane en 4(20), 5. Tetrahedron, **47**, 9823 (1991).

72. EYRE, E.D., J.W. HARRISON, and B. LYTHGOE: Taxine. Part VI. The Stereochemistry of Taxicin-I and Taxicin-II. J. Chem. Soc. C 452 (1967).

73. FALZONE, J.J., A.J. BENESI, and J.T.J. LECOMTE: Characterization of Taxol in Methylene Chloride by NMR Spectroscopy. Tetrahedron Letters **33**, 1169 (1992).

74. FÉTIZON, M.I. HANNA, and R. ZEGHDOUDI: Synthetic Studies on Taxane Diterpenes. II. Preparation of a Key Intermediate: 7-exo-Benzoyloxy-9,9-Dimethyl Bicyclo-[3.3.1]Nonan-2,4-Diene. Syn. Comm. **16**, 1 (1986).

75. FILLION, H.: 5-Hydroxy-1,8-naphthalene Carbolactone. Esterification by the Taxol "Side Chain". Pharmazie **45**, 287 (1990).

76. FREJD, T., and L. PETTERSSON: Synthetic Studies Toward Taxol; the A,C-ring System. 8th International IUPAC Conference on Organic Synthesis, Helsinki, Finland, 1990.

77. FREJD, T., G. MAGNUSSON, and L. PETTERSSON: Efforts Towards an Enantiospecific Synthesis of the Taxol A-Ring. Chemica Scripta **27**, 561 (1987).

78. FUNK, R.L., W.J. DAILY, and M. PARVEZ: A Convergent Approach to the Taxane Class of Compounds. J. Org. Chem. **53**, 4141 (1988).

79. FURUKAWA, T., Y. HORIGUCHI, and I. KUWAJIMA: Synthetic Studies of Taxusin. 32nd Symposium on the Chemistry of Natural Products, Chiba, 1990.

80. GADWOOD, R.C., and R.M. LETT: Preparation and Rearrangement of 1,2-Dialkenylcyclobutanols. A Useful Method for Synthesis of Substituted Cyclooctenones. J. Org. Chem. **47**, 2268 (1982).

81. GRAF, E.: Taxin B, das Hauptalkaloid von *Taxus baccata* L. Arch. Pharm. 443 (1958).
82. GRAF, E. and H. BERTHOLDT: Das amorphe Taxin und das Kristallisierte Taxin A. Pharmazeutische Zentralhalle **96**, 385 (1957).
83. GRAF, E. A. KIRFEL, G.-J. WOLFF, and E. BREITMAIER: Die Aufklärung von Taxin A aus *Taxus baccata* L. Liebigs Ann. Chem. 376 (1982).
84. GRAF, E., S. WEINANDY, B. KOCH, and E. BREITMAIER: ^{13}C-NMR-Untersuchung von Taxin B aus *Taxus baccata* L. Leibigs Ann. Chem. 1147 (1986).
85. GREM, J.L., K.D. TUTSCH, K.J. SIMON, J.K.V. WILLSON, D.C. TORMEY, S. SWAMIN-ATHAN, and D.L. TRUMP: Phase I Study of Taxol Administered as a Short IV Infusion Daily for 5 Days. Cancer Treat. Rep. **71**, 1179 (1987).
86. GU, J., R. ZHENG, Z. ZHANG, and Z. JIA: Inhibition of Taxanes on DNA and Protein Syntheses of Tumor Cells. Planta Med. **57**, 495 (1991).
87. GUÉRITTE-VOEGELEIN, F., D. GUÉNARD, and P. POTIER: Taxol and Derivatives: A Biogenetic Hypothesis. J. Nat. Prod. **50**, 9 (1987).
88. GUÉRITTE-VOEGELEIN, F., D. GUÉNARD, F. LAVELLE, M.-T. LE GOFF, L. MANGATAL, and P. POTIER: Relationships Between the Structure of Taxol Analogues and Their Antimitotic Activity. J. Med. Chem. **34**, 992 (1991).
89. GUÉRITTE-VOEGELEIN, F., V. SENILH, B. DAVID., D. GUÉNARD, and P. POTIER: Chemical Studies of 10-Deacetyl Baccatin III. Hemisynthesis of Taxol Derivatives. Tetrahedron **42**, 4451 (1986).
90. GUÉRITTE-VOEGELEIN, F., D. GUÉNARD, L. MANGATAL, P. POTIER, J. GUILHEM, M. CESARIO, and C. PASCARD: Structure of a Synthetic Taxol Precursor: N-*tert*-Butoxycarbonyl-10-deacetyl-N-debenzoyltaxol. Acta Cryst. **C46**, 781 (1990).
91. GUPTA, R.S., and A.K. DUDANI: Mechanism of Action of Antimitotic Drugs: a New Hypothesis Based on the Role of Cellular Calcium. Med. Hypotheses **28**, 57 (1989).
92. HAMEL, E., A.A. DEL CAMPO, M.C. LOWE, and C.M. LIN: Interactions of Taxol, Microtubule-associated Proteins, and Guanine Nucleotides in Tubulin Polymerization. J. Biol. Chem. **256**, 11887 (1981).
93. HARADA, K., and Y. NAKAJIMA: Optical Resolution and Configuration of *cis*-β-Phenylglycidic Acid. Bull. Chem. Soc. Jpn. **47**, 2911 (1974).
94. HARADA, N., and K. NAKANISHI: A Method for Determining the Chiralities of Optically Active Glycols. J. Am. Chem. Soc. **91**, 3989 (1969).
95. HARADA, N., M. OHASHI, and K. NAKANISHI: The Benzoate Sector Rule, a Method for Determining the Absolute Configurations of Cyclic Secondary Alcohols. J. Am. Chem. Soc. **90**, 7349 (1968).
96. HARRISON, J.W., and B. LYTHGOE: Taxine. Part III. A Revised Structure for the Neutral Fragment from O-Cinnamoyltaxicin-I. J. Chem. Soc. C, 1932 (1966).
97. HARRISON, J.W., R.M. SCROWSTON, and B. LYTHGOE: Taxine. Part IV. The Constitution of Taxine-I. J. Chem. Soc. C, 1933 (1966).
98. HARTZELL Jr., H.: The Yew Tree: a Thousand Whispers. Eugene, OR: Hulogosi 1991.
99. HATANAKA, Y., and I. KUWAJIMA: Reactions of 3[(Trimethylsilyl)methyl]cyclo-2-hexenone with Carbonyl Compounds. Regio- and Chemoselective Condensations. J. Org. Chem. **51**, 1932 (1986).
100. HAYAKAWA, K., S. OHSUKI, and K. KANEMATSU: A Highly Efficient Synthesis of Bicyclo[n.3.1]Ring Systems by Allene Intramolecular Cycloaddition: Tandem Intramolecular [2+2]Cycloaddition-[3,3]-Sigmatropic Rearrangement. Tetrahedron Letters **27**, 4205 (1986).
101. HENRY, T.A.: The Plant Alkaloids, 4th edition, p. 770. Philadelphia: The Blakiston Company 1949.

102. Ho, T.-I., G.-H. Lee, S.-M. Peng, M.-K. Yeh, F.-C. Chen, and W.L. Yang: Structure of Taxusin. Acta Cryst. C43, 1378 (1987).
103. Ho, T.-I., Y.-C. Lin, G.-H. Lee, S.-M. Peng, M.-K. Yeh, and F.-C. Chen: Structure of Taiwanxan. Acta Cryst. C43, 1380 (1987).
104. Hoke, S.H. III, J.M. Wood, R.G. Cooks, H. Jayasuria, P. Heinstein, C.-J. Chang, S.W. Brobst, N. Wheeler, and K.M. Snader: Tandem Mass Spectrometric Analysis of Taxanes from Taxus brevifolia. American Society of Pharmacognosy 32nd Annual Meeting, Chicago, IL, July 1991, Abstract P68.
105. Holmes, F.A., R.S. Walters, R.L. Thriault, A.D. Forman, L.K. Newton, M.N. Raber, A.U. Buzdar, D.K. Frye, and G.N. Hortobagyi: Phase II Trial of Taxol, an Active Drug in the Treatment of Metastatic Breast Cancer. J. Nat. Cancer Inst. 83, 1797 (1991).
106. Holton, R.A.: Synthesis of the Taxane Ring System. J. Am. Chem. Soc. 106, 5731 (1984).
107. Holton, R.A.: Approaches to the Total Synthesis of Taxol. In: Workshop on Taxol and Taxus: Current and Future Perspectives. National Cancer Institute, Bethesda, MD, June 26, 1990.
108. Holton, R.A.: Method for Preparation of Taxol Using an Oxazinone. U.S. Patent 5,015,744, May 14, 1991.
109. Holton, R.A., R.R. Juo, H.B. Kim, A.D. Williams, S. Harusawa, R.E. Lowenthal, and S. Yogai: A Synthesis of Taxusin. J. Am. Chem. Soc. 110, 6558 (1988).
110. Hönig, H., P. Seufer-Wasserthal, and H. Weber: Chemo-enzymatic Synthesis of all Isomeric 3-Phenylserines and -Isoserines. Tetrahedron 46, 3841 (1990).
111. Horiguchi, Y., T. Furukawa, and I. Kuwajima: A Highly Efficient Eight-Membered-Ring Cyclization for Construction of the Taxane Carbon Framework. J. Am. Chem. Soc. 111, 8277 (1989).
112. Horwitz, S.B., L. Lothstein, J.J. Manfredi, W. Mellado, J. Parness, S.N. Roy, P.B. Schiff, L. Sorbara, and R. Zeheb: Taxol: Mechanisms of Action and Resistance. Ann. N.Y. Acad. Sci. 466, 733 (1986).
113. Huang, C.H.O., D.G.I. Kingston, N.F. Magri, and G. Samaranayake: New Taxanes from Taxus brevifolia, 2. J. Nat. Prod. 49, 665 (1986).
114. Huttel, M.S., A.S. Olesen, and E. Stoffersen: Complement-Mediated Reactions to Diazepam with Cremophor as Solvent (Stesolic MR). Br. J. Anaesth. 52, 77 (1980).
115. Inouye, Y.: Personal communication (1991).
116. Inouye, Y., C. Fukaya, and H. Kaksawa: Preparation of Several 1,5,5,11,11-Pentamethyl-tricyclo[6.2.1.0^{2,6}]undecane Derivatives. Bull. Chem. Soc. Jpn. 54, 1117 (1981).
117. Inouye, Y., Kojima, J. Owada, and H. Kakisawa: Preparation of Bicyclo[3.3.1] nonane-2,4-dione Derivatives. Bull. Chem. Soc. Jpn. 60, 4369 (1987).
118. IUPAC: Nomenclature of Organic Chemistry, Section F: Natural Products and Related Compounds. Eur. J. Biochem. 86, 1 (1978).
119. Jackson, C.B., and G. Pattenden: Total Synthesis of Verticillene, the Putative Biogenetic Precursor of the Taxane Alkaloids. Tetrahedron Letters 26, 3393 (1985).
120. Jaziri, M., B.M. Diallo, M.H. Vanhaelen, R.J. Vanhaelen-Fastre, A. Zhiri, A.G. Gecu, and J. Homes: Enzyme-linked immunosorbent Assay for the Detection and the Semi-quantitative Determination of Taxane Diterpenoids Related to Taxol in Taxus sp. and Tissue cultures. J. Pharm. Belg. 46, 93 (1991).
121. Jennings, D.W., H.M. Deutsch, L.H. Zalkow, and A.S. Teja: Supercritical Extraction of Taxol from the Bark of Taxus brevifolia. J. Supercrit. Fluids. 5, 1 (1992).

122. JIA, Z.-J., and Z.-P. ZHANG: Taxanes from *Taxus chinensis*. Chin. Sci. Bull. **36**, 1174 (1991).

123. JITRANGSRI, C.: Approaches to the Synthesis of Modified Taxols. Ph.D. Dissertation. Virginia Polytechnic Institute and State University, Blacksburg, 1986.

124. KACZMAREK, R., and S. BLECHERT: De-Mayo-Reaktionen Mit Allen. Ein Kurzer Weg Zum Bicyclo[5.3.1]undecansystem Der Taxane. Tetrahedron Letters **27**, 2845 (1986).

125. KAJI, E., A. IGARASHI, and S. ZEN: The Synthetic Reactions of Aliphatic Nitro Compounds. XI. The Synthesis of β-Amino-α-hydroxycarboxylic Acids and γ-Amino-carboxylic Acids. Bull. Chem. Soc. Jpn. **49**, 3181 (1976).

126. KAMANDI, K., A.W. FRAHM, and F. ZYMALKOWSKI: β-Phenyl-isoserines by Ammonolysis of β-Phenylglycidates. Arch. Pharmaz. **307**, 871 (1974).

127. KARLSSON, B., A.-M. PILOTTI, A.-C. SODERHOLM, T. NORIN, S. SUNDIN, and M. SUMIMOTO. The Structure and Absolute Configuration of Verticillol, a Macrocyclic Diterpene Alcohol from the Wood of *Sciadopitys verticillata* Sieb. et Zuzz. (Taxodiaceae). Tetrahedron **34**, 2349 (1978).

128. KATO, T., H. TAKAYANAGI, T. SUZUKI, and T. UYEHARA: Cyclization of Polyenes XXIX. Synthesis of Seco-Taxane Skeleton Based on Biogenetical Consideration. Tetrahedron Letters **14**, 1201 (1978).

129. KENDE, A.S., S. JOHNSON, P. SANFILIPPO, J.C. HODGES, and L.N. JUNGHEIM: Synthesis of a Taxane Triene. J. Am. Chem. Soc. **108**, 3513 (1986).

130. KHAN, N.U.-D., and N. PARVEEN: The Constituents of the Genus *Taxus*. J. Sci. Ind. Res. **46**, 512 (1987).

131. KINGSTON, D.G.I.: The Chemistry of Taxol. Pharmac. Ther. **52**, 1 (1991).

132. KINGSTON, D.G.I., D.R. HAWKINS, and L. OVINGTON: New Taxanes from *Taxus brevifolia*. J. Nat Prod. **45**, 466 (1982).

133. KINGSTON, D.G.I., A.A.L. GUNATILAKA, and C.A. IVEY: Modified Taxols, 7. A Method for the Separation of Taxol and Cephalomannine. J. Nat. Prod. **55**, 259 (1992).

134. KINGSTON, D.G.I., N.F. MAGRI, and C. JITRANGSRI: Synthesis and Structure-Activity Relationships of Taxol Derivatives as Anticancer Agents. In: New Trends in Natural Products Chemistry. Eds ATTA-UR-RAHMAN, and P.W. LEQUESNE, p. 219. Amsterdam. Elsevier Science Publishers, 1986.

135. KINGSTON, D.G.I., G. SAMARANAYAKE, and C.A. IVEY: The Chemistry of Taxol, a Clinically Useful Anticancer Agent. J. Nat. Prod. **53**, 1 (1990).

136. KINGSTON, D.G.I., and Z,-Y ZHAO: Water Soluble Derivatives of Taxol. U.S. Patent 5,059,699, Oct. 22, 1991.

137. KITAGAWA, I., H. SHIBUYA, H. FUJIOKA, A. KAJIWARA, Y. YAMAMOTO, S. TSUJII, and K. MURAKAWA: Chemical Transformations of Terpenoids II. Acid Treatment of (3S)-1-Vinyl,(3S)-1-Hydroxy-propenyl-, and (3S)-1-Epoxyethnyl-1,2,2-trimethyl-cyclopentane Derivatives: Ring Enlargement Reactions and Successive Migrations of Methyl Residues. Chem. Pharm. Bull. **29**, 2548 (1981).

138. KITAGAWA, I., H. SHIBUYA, H. FUSIOKA, Y. YAMAMOTO, A. KAJIWARA, K. KITAMURA, and A. MIYAO: Successive Methyl Migration Occurring in the Acid Treatment of 1-Epoxyethyl-1,2,2,-Trimethyl-Cyclopentane Derivative. Tetrahedron Letters **21**, 1963 (1980).

139. KITAGAWA, I., H. SHIBUYA, H. FUJIOKA, A. KAJIWARA, S. TSJUII, Y. YAMAMOTO, and A. TAKAGI: Synthesis of a Chiral 1,1,3-Trimethylcyclohexane Derivative from d-Camphor: A Potent Key Building Block for ENT-Taxane-Type Diterpenoids. Chemistry Letters 1001 (1980).

140. KITAGAWA, I., H. SHIBUYA, H. FUJIOKA, A. KAJIWARA, Y. YAMAMOTO, S.TSUJII, A. TAKAGI, and M. HORI: Chemical Transformation of Terpenoids I, Syntheses of (3S)-1-

Vinyl-,(3S)-1-Hydroxy-propenyl-, and (3S)-1-Epoxyethyl-1,2,2-trimethylcyclopentane
Derivatives from d-Camphor via d-Camphoric Acid. Chem. Pharm. Bull. **29**, 2540
(1981).

141. KITAGAWA, I., H. SHIBUYA, H. FUJIOKA, A. KAJIWAKA, Y. YAMAMOTO, A. TAKAGAI, K.
SUZUKI, and M. HORI: Tennen Yuki Kagobutsu Toronkai Koen Yoshishu **22**, 132
(1979). Chemical Abstracts has evidently abstracted this report twice: Chem. Abstr. **92**,
198568g (1980); Chem. Abstr. **93**, 168439u (1980).

142. KITAGAWA, I., S. TSUJII, H. FUJIOKA, A. KAJIWARA, Y. YAMAMOTO, and H. SHIBUYA:
Chemical Transformations of Terpenoids VI. Syntheses of Chiral Segments, Key
Building-Blocks for the Left Half of Taxane-Type Diterpenoids. Chem. Pharm. Bull.
32, 1294 (1984).

143. KOBAYASHI, T., M. KURONO, H. SATO, and K. NAKANISHI: Nature of Photochemically
Induced Transannular Hydrogen Abstractions of Taxinines. J. Am. Chem. Soc. **94**,
2863 (1972).

144. KOJIMA, T., Y. INOUYE, and H. KAKISAWA: Synthesis of a ($\pm$)-3β-Trinortaxane
Derivative. Chem. Letters 323 (1985).

145. KOJIMA, T., Y. INOUYE, and H. KAKISAWA: 52nd Annual Meeting of the Chemical
Society of Japan (Kyoto, 1986); Abstract 1N34.

146. KONDO, H., and T. TAKAHASHI: Constituents of the Yew Leaves. II. J. Pharm. Soc.
Jpn. **524**, 861 (1925).

147. KRAUS, G.A., P.J. THOMAS, and Y. HON: The Reaction of 1,1,-Dimethoxyethylene
with Bridgehead Enones. A Direct Approach to the Taxane Skeleton. J. Chem. Soc.
Chem. Commun. 1849 (1987).

148. KRIS, M.S., J.P. O'CONNELL, R.J. GRALLA, M.S. WERTHEIM, R.M. PARENTE, P.B.
SCHIFF, and C.W. YOUNG: Phase I Trial of Taxol Given as 3-Hour Infusion Every 21
Days. Cancer Treat. Rep. **70**, 605 (1986).

149. KUMAGAI, T., F. ISE, T. UYEHARA, and T. KATO: Synthesis of *Trans*-4, 8, 12, 15, 15-
pentamethylbicyclo[9.3.1]pentadeca-3E,7Z,12Z-triene, a Geometrical Isomer of Anhydroverticillol. Chem. Letters 25 (1981).

150. KURONO, M.: Structure of Taxinine. D.Sc. Dissertation, Nagoya University, Nagoya,
Japan, 1959.

151. KURONO, M., Y. NAKADAIRA, S. ONUMA, K. SASAKI and K. NAKANISHI: Taxinine.
Tetrahedron Letters 2153 (1963).

152. KURONO, M., Y. MAKI, K. NAKANISHI, M. OHASHI, K. UEDA, S. UYEO, M.C. WOODS,
and Y. YAMAMOTO. The Stereochemistry of Taxinine. Tetrahedron Letters 1917
(1965).

153. KUWAJIMA, I: Kagaku to Kogyo. Toyko: Nippon Kagakkai **44**, 662 (1991).

154. LEE, C.L., Y. HIROSE, and T. NAKATSUKA: Inhibitory Effect of *Taxus mairei* Heartwood Extractives on the Curing of Unsaturated Polyester Resins. Mokuzaki
Gakkaishi **21**, 249 (1975).

155. LEGHA, S.S., S. RING, N. PAPADOPOULOS, M. RABER, and R.S. BENJAMIN: A Phase II
Trial of Taxol in Metastatic Melanoma. Cancer **65**, 2478 (1990).

156. LEGHA, S.S., D.M. TENNEY, and I.R. KRAKOFF: Phase I Study of Taxol Using a 5-Day
Intermittent Schedule. J. Clin. Oncol. **4**, 762 (1986).

157. LEETE, E., and G.B. BODEM: The Biosynthesis of 3-Dimethylamino-3-Phenyl-
Propanoic Acid in Yew. Tetrahedron Letters **33**, 3925 (1966).

158. LIAN, J.-Y., Z.-D. MIN, M. MIZUNO, T. TANAKA, and M. IINUMA: Two Taxane
Diterpenes from *Taxus mairei*. Phytochem. **27**, 3674 (1988).

159. LIANG, J.-Y., Z.-D. MIN, and M. NIWA: Studies on the Diterpenes of *Taxus mairei*. II.
Structure of 2-deacetoxy-taxinine. J. Acta Chim. Sin (Huaxe Xuebao) **46**, 1053 (1988).

160. LIN, J., M.N. NIKAIDO, and G. CLARK: Convergence with Mutual Kinetic Resolution

1. Studies Defining Methodology for the Taxol C/D Ring Fragment and Synthesis of the A Ring Fragment. J. Org. Chem. **52**, 3745 (1987).

161. LIU, C.-L., Y.-C. LIN, Y.-M. LIN, and F.-C. CHEN: Constituents of the Heartwood of Taiwan Yew. Tai'wan Ko'hsueh **38**, 119 (1984).

162. LONGNECKER, S.M., R.C. DONEHOWER, A.E. GATES, T.L. CHEN, R.B. BRUNDRETT, L.B. GROCHOW and D.S. ETTINGER: High-performance Liquid Chromatographic Assay for Taxol in Human Plasma and Urine and Pharmacokinetics in a Phase I Trial. Cancer Treat. Rep **71**, 53 (1987).

163. LUCAS, H.: Über ein in Den Blättern von *Taxus baccata* L. enthaltenes Alkaloid (das Taxin). Archiv der Pharmazie **85**, 145 (1856).

164. LYTHGOE, B.: The Taxus Alkaloids In: The Alkaloids, Vol 10, Chemistry and Physiology, Ed. R.H.F. MANSKE, p. 597. New York: Academic Press. 1968.

165. LYTHGOE, B., K. NAKANISHI, and S. UYEO: Taxane. Proc. Chem. Soc., 301 (1964).

166. MAGRI, N.F.: Modified Taxols as Anticancer Agents. Ph.D. Dissertation, Virginia Polytechnic Institute and State University, Blacksburg, 1985.

167. MAGRI, N.F., and D.G.I. KINGSTON: Modified Taxols. 2. Oxidation Products of Taxol. J. Org. Chem. **51**, 797 (1986).

168. MAGRI, N.F., D.G.I. KINGSTON, C. JITRANGSRI, and T. PICCARIELLO: Modified Taxols. 3. Preparation and Acylation of Baccatin III. J. Org. Chem. **51**, 3239 (1986).

169. MAGRI, N.F., and D.G.I. KINGSTON: Modified Taxols. 4. Synthesis and Biological Activity of Taxols Modified in the Side Chain. J. Nat. Prod. **51**, 298 (1988).

170. MAKI, Y., and K. YAMANE: Taxinine Derivatives. The Stereochemistry of Isopropylidene Dihydrotaxinolactone, a Novel Autooxidation Product of Isopropylidenedihydrotaxinol. Chem. Pharm. Bull. **17**, 2071 (1969).

171. MANFREDI, J.J., and S.B. HORWITZ: TAXOL: An Antimitotic Agent with a New Mechanism of Action. Pharmac. Ther. **25**, 83 (1984).

172. MANGATAL, L., M.-T. ADELINE, D. GUÉNARD, F. GUÉRITTE-VOEGELEIN, and P. POTIER. Application of the Vicinal Oxyamination Reaction with Asymmetric Induction to the Hemisynthesis of Taxol and Analogues. Tetrahedron **45**, 4177 (1989).

173. MARTIN, S., J.B. WHITE, and R. WAGNER: Alkoxide-Accelerated Sigmatropic Rearrangements. A Novel Entry to the Bicyclo[5.3.1]undec-7ene System of the Taxane Diterpenes. J. Org. Chem. **47**, 3190 (1982).

174. MATHEW, A.E., M.R. MEJILLANO, J.P. NATH, R.H. HIMES, and V.J. STELLA: Synthesis and Evaluation of Some Water-Soluble Prodrugs and Derivatives of Taxol with Antitumor Activity. J. Med. Chem. **35**, 145 (1992).

175. MCLAUGHLIN, J.L., R.W. MILLER, R.G. POWELL, and C.R. SMITH, Jr.: 19-Hydroxybaccatin-III, 10-Deacetylcephalomannine, and 10-Deacetyltaxol: New Antitumor Taxanes from *Taxus wallichiana*. J. Nat. Prod. **44**, 312 (1981).

176. MCCLURE, T.D., M.L.J. REIMER, and K.H. SCHRAM: Mass Spectrometry of Taxol and Taxol Analogs. The 38th ASMS Conference on Mass Spectrometry and Allied Topics, WP 135 (1990).

177. MCGUIRE, W.P., E.K. ROWINSKY, N.B. ROSENSHEIN, F.C. GRUMBINE, D.S. ETTINGER, D.K. ARMSTRONG, and R.C. DONEHOWER: Taxol: A Unique Antineoplastic Agent with Significant Activity in Advanced Ovarian Epithelial Neoplasms. Ann. Internal Med. **111**, 273 (1989).

178. MELLADO, W., N.F. MAGRI, D.G.I. KINGSTON, R. GARCIA-ARENAS, G.A. ORR, and S.B. HORWITZ: Preparation and Biological Activity of Taxol Acetates. Biochem. Biophys. Res. Commun. **124**, 329 (1984).

179. METACHEM TECHNOLOGIES, INC.: Taxol Analysis. MetaChem Catalog, 27 (1992).

180. MILLER, R.W.: A Brief Survey of Taxus Alkaloids and Other Taxane Derivatives. J. Nat. Prod. **43**, 425 (1980).

181. MILLER, R.W., R.G. POWELL, C.R. SMITH Jr., E. ARNOLD, and J. CLARDY: Antileukemic Alkaloids from *Taxus wallichiana* Zucc. J. Org. Chem. **46**, 1469 (1981).

182. MIN, Z.-D., H. JIANG, and J.-Y LIANG: Studies on the Taxane Diterpenes of the Heartwood from *Taxus mairei*. Acta Pharm. Sin. (Yaoxue Xuebao) **24**, 673 (1989).

183. MIYAZAKI, M., K. SHIMIZU, H. MISHIMA, and M. KURABAYASHI: The Constituent of the Heartwood of *Taxus cuspidata* Sieb. et. Zucc. Chem. Pharm. Bull. **16**, 546 (1968).

184. MONSARRAT, B., E. MARIEL, S. CROS, M. GARES, D. GUÉNARD, F. GUÉRITTE-VOEGELEIN, and M. WRIGHT: Taxol Metabolism. Isolation and Identification of Three Major Metabolites of Taxol in Rat Bile. Drug Metab. Disp. **18**, 895 (1990).

185. MORELLI, I.: Constituenti di *Taxus baccata* L. Fitoterapia **47**, 31 (1976).

186. NAGAOKA, H., K. OHSAWA, T. TAKATA, and Y. YAMADA: A New Synthetic Approach to the Bicyclo[5.3.1]undecane Ring System in Taxanes. Tetrahedron Letters **25**, 5389 (1984).

187. NAKANISHI, K., and M. KURONO: Some NMDR Studies on Taxinine and Derivatives. Tetrahedron Letters **39**, 2161 (1963).

188. NATIONAL CANCER INSTITUTE, DIVISION OF CANCER TREATMENT: Taxol (IND 22850, NSC 125973). Annu. Rep. to FDA, March (1991).

189. NEH, H., S. BLECHERT, W. SCHNICK, and M. JANSEN: A Novel Entry to the Taxane Structural Unit. Angew. Chem. Int. Ed. Engl. **23**, 905 (1984).

190. NEH, H., A. KÜHLING, and S. BLECHERT: 12, Stereoselektive Synthesen von Taxane-Derivaten. Helv. Chim Acta **72**, 101 (1989).

191. NETO, A.G.F., and F. DICOSMO: Distribution and Amounts of Taxol in Different Shoot Parts of *Taxus cuspidata* Siebold and Zucc. Male and Female Plants. Planta Med. **58**, xxx (1992).

192. OHNUMA, T., A.S. ZIMET, V.A. COFFEY, J.F. HOLLAND, and E.M. GREENSPAN: Phase I Study of Taxol in a 24-Hr. Infusion Schedule. Proc. Am. Assoc. Cancer Res. **26**, 167 (1985).

193. OHTUSKA, Y., and T. OISHI: An Effective Synthesis of Medium-Ring Ketones. Tetrahedron Letters **46**, 4487 (1979).

194. OHTUSKA, Y., and T. OISHI: Medium-Ring Ketone Synthesis, Intramolecular Acylation of Sulfur-stabilized Carbanions: A Model Study. Chem. Pharm. Bull. **31**, 443 (1983).

195. OHTSUKA, Y., and T. OISHI: Medium-Ring Ketone Synthesis. Synthesis of Eight-to-Twelve-membered Cyclic Ketones Based on the Intramolecular Cyclization of Large-membered Lactam Sulfoxides or Sulfones. Chem. Pharm. Bull. **31**, 454 (1983).

196. OHTSUKA, Y., and T. OISHI: An Approach to Taxane Diterpenes. The Synthesis of the Bicyclo[5.3.1]undecane-4-one Derivatives. Heterocycles **21**, 371 (1984).

197. OHTSUKA, Y., and T. OISHI: A Synthetic Approach to Taxane Diterpenes. A Synthesis of the Bicyclo[5.3.1]undecenone Ring System. Tetrahedron Letters **27**, 203 (1986).

198. OHTSUKA, Y., and T. OISHI: RIKEN Report, p. 102–104, **63**, 97 (1987).

199. OHTSUKA, Y., and T. OISHI: Studies on Taxane Synthesis. I. Synthesis of 3,8,11,11-Tetramethyl-4-oxobicyclo[5.3.1]undecane as a Model for Taxane Synthesis. Chem. Pharm. Bull **36**, 4711 (1988).

200. OHTSUKA, Y., and T. OISHI: Studies on Taxane Synthesis. II. Synthesis of 3,8,11,11-Tetramethyl-4-oxo- and 4,8,11,11-Tetramethyl-3-oxo[5.3.1]undec-8-enes Corresponding to the A- and B-Rings of Taxane Diterpenes. Chem. Pharm. Bull. **36**, 4722 (1988).

201. OHTSUKA, Y., and T. OISHI: 109th Annual Meeting of Pharm. Soc. of Japan. Nagoya, 1989. Abstracts of Papers III, p. 88.
202. OHTSUKA, Y., and T. OISHI: Studies on Taxane Synthesis. III. Stereocontrolled synthesis of a Twelve-Membered Lactam Sulfide as a Precursor of 4,8,11,11-Tetramethyl-3-oxobicyclo[5.3.1]undec-8-ene. Chem. Pharm. Bull. **39**, 1359 (1991).
203. OISHI, T., and Y. OTSUKA: Synthetic Studies on Natural Products Possessing Eight- or Nine-membered Rings. In: Studies in Natural Products Chemistry, Ed. Atta-ur-Rahman, Vol. 111, p. 73. Amsterdam: Elsevier Science Publishers B.V. 1989.
204. OJIMA, I., I. HABUS, M. ZHAO, G.I. GEORG and L.R. JAYASINGHE: Efficient and Practical Asymmetric Synthesis of the Taxol C-13 Side Chain, N-Benzoyl-(2R, 3S)-3-phenylisoserine, and its Analogues via Chiral 3-Hydroxy-4-aryl-β-lactams through Chiral Ester Enolate-Imine cyclocondensation. J. Org. Chem. **56**, 1681 (1991).
205. PALOMO, C., A. ARRIETA, F.P. COSSIO, J.M. AIZPURUA, A. MIELGO, and N. AURREK-OETXEA: Highly Stereoselective Synthesis of α-Hydroxy β-Amino Acids through β-Lactams: Application to the Synthesis of the Taxol and Bestatin Side Chains and Related Systems. Tetrahedron Letters **31**, 6429 (1990).
206. PARNESS, J., and S.B. HORWITZ: Taxol Binds of Polymerized Tubulin *in vitro*. J. Cell. Biol. **91**, 479 (1981).
207. PAQUETTE, L.A.: Studies Directed Toward the Total Synthesis of the Taxanes. In: Studies in Natural Products Chemistry, Vol. 10. Ed. Atta-ur-Rahman. p.xxx. Amsterdam: Elsevier. 1992.
208. PAQUETTE, L.A., D.T. DeRUSSY, N.A. PEG, R.T. TAYLOR, and T.M. ZYDOWSKY: Direct Oxygenation of Enolates Generated by Anionic Oxy-Cope Rearrangement. Expedient Preparation of Polycyclic α-Hydroxy Ketones. J. Org. Chem. **54**, 4576 (1989).
209. PAQUETTE, L.A., N.A. PEGG, D. TROOPS, G.D. MAYNARD, and R.D. ROGERS: [3.3] Sigmatropy within 1-Vinyl-2-alkenyl-7,7-dimethyl-*exo*-norbornan-2-ds. The First Atropselective Oxyanionic Cope Rearrangement. J. Am. Chem. Soc. **112**, 277 (1990).
210. PETERSON, J.R., H.D. DO. and R.D. ROGERS: X-ray Structure and Crystal Lattice Interactions of the Taxol Side-Chain Methyl Ester. Pharm. Res. **8**, 908 (1991).
211. PETTERSSON, L., T. FREJD, and G. MAGNUSSON: An Enantiospecific Synthesis of a Taxol A-Ring Building Unit. Tetrahedron Letters **28**, 2753 (1987).
212. PLATT, R.V., C.T. OPIE, and E. HASLAM: Biosynthesis of Flavan-3-ols and Other Secondary Plant Products from (2S)-Phenylalanine. Phytochem. **23**, 2211 (1984).
213. POWELL, R.G., R.W. MILLER, and C.R. SMITH, JR.: Cephalomannine: a New Antitumor Alkaloid from *Cephalotaxus mannii*. J. Chem. Soc. Chem. Commun. 102 (1979).
214. PRELOG, V., P. BARMAN, and M. ZIMMERMAN: Weitere Untersuchungen über die Gültigkeitsgrenzen der Bredt' schen Regel. Eine Variante der Robinson's chem. Synthese von Cyclischen Ungesättigten Ketonen. Helv. Chim Acta **32**, 1284 (1949).
215. RAO, C., X. LIU, P.L. ZHANG, W.M. CHEN, and Q.C. FANG: Application of High-speed Countercurrent Chromatography for the Isolation of Natural Products and Preparative Isolation of Taxane Diterpenoids and Diterpene Alkaloids. Acta Pharm. Sin. (Yaoxue Xuebao) **26**, 510 (1991).
216. RINGEL, I., and S.B. HORWITZ: Taxol is Converted to 7-Epitaxol, A Biologically Active Isomer, in Cell Culture Medium. J. Pharmacol. Exp Ther. **242**, 692 (1987).
217. RINGEL, I., and S.B. HORWITZ: Studies with RP56976 (Taxotere): a Semisynthetic Analogue of Taxol. J. Nat. Cancer Inst. **83**, 288 (1991).
218. RIZZO, J., C.RILEY, D. VON HOFF, J. KUHN, J. PHILLIPS, and T. BROWN: Analysis of Anticancer Drugs in Biological Fluids: Determination of Taxol with Application to Clinical Pharmacokinetics. J. Pharm. Biomed. Anal. **8**, 159 (1990).
219. ROJAS, A.C., D. DE MARCANO, B. MENDEZ, and J. DE MENDEZ: Carbon-13 NMR

Spectra of Taxane-Type Diterpenes: Oxiranes and Oxetanes. Org. Magn. Reson. **21**, 257 (1983).

220. ROWINSKY, E.K., P.J. BURKE, J.E. KARP, D.S. ETTINGER, R.W. TUCKER, and R.C. DONEHOWER: Phase I Study of Taxol in Refractory Adult Acute Leukemia. Proc. Am. Assoc. Cancer Res. **29**, 215 (1988).

221. ROWINSKY, E.K., P.J. BURKE, J.E. KARP, R.W. TUCKER, D.S. ETTINGER, and R.C. DONEHOWER: Phase I and Pharmacodynamic Study of Taxol in Refractory Acute Leukemias. Cancer Res. **49**, 4640 (1989).

222. ROWINSKY, E.K., L.A. CAZENAVE, and R.C. DONEHOWER: Taxol: A Novel Investigational Antimicrotubule Agent. J. Nat. Cancer Inst. **82**, 1247 (1990).

223. ROWINSKY, E.K., and R.C. DONEHOWER: Taxol: Twenty Years Later, the Story Unfolds. J. Nat. Cancer Inst. **83**, 1778 (1991).

224. ROWINSKY, E.K., W.P. MCGUIRE, T. GUARINERI, J.S. FISHERMAN, M.C. CHRISTIAN, and R.C. DONEHOWER: Cardiac Disturbances During the Administration of Taxol. J. Clin. Oncol. **9**, 1704 (1991).

225. SAHA, G., A. BHATTACHARYA, S.S. ROY, and S. GHOSH: A Simple Approach to the Construction of Bicyclo[5.2.1] and [5.3.1] Ring Systems from Bicyclo[2.2.1]heptane Precursors. Tetrahedron Letters **31**, 1483 (1990).

226. SAHA, G., and S. GHOSH: A New Route to the Synthesis of 7-Functionalized Bicyclo [2.2.1]heptane Derivatives. Syn. Commun. **21**, 2129 (1991).

227. SAKAN, K.D., and B.M. CRAVEN: Synthetic Studies on the Taxane Diterpenes. Utility of the Intramolecular Diels-Alder Reaction for a Single-Step Stereocontrolled Synthesis of a Taxane Model System. J. Am. Chem. Soc. **105**, 3732 (1983).

228. SAKAN, K., D.A. SMITH, S.A. BABIRAD, F.R. FRONCZEK, and K.N. HOUK: Stereoselectivities of Intramolecular Diels-Alder Reactions. Formation of the Taxane Skeleton. J. Org. Chem. **56**, 2311 (1991).

229. SAMARANAYAKE, G.: Studies on the Chemistry of Taxol. Ph.D. Dissertation. Virginia Polytechnic Institute and State University, Blacksburg, 1990.

230. SAMARANAYAKE, G., N.F. MAGRI, C. JITRANGSRI, and D.G.I. KINGSTON: Modified Taxols. 5. Reaction of Taxol with Electrophilic Reagents and Preparation of a Rearranged Taxol Derivative with Tubulin Assembly Activity. J. Org. Chem. **56**, 5114 (1991).

231. SCHELL, F.M., P.M. COOK, S.W. HAWKINSON, R.E. CASSAOY, and W.E. THIESSEN: Intramolecular Photochemical Cycloaddition of a Vinylogous Imide. Crystals and Molecular Structure of a Tetracyclic $C_{17}H_{25}O_2N$ Product. J. Org. Chem. **44**, 1380 (1979).

232. SCHIFF, P.B., J. FANT and S.B. HORWITZ: Promotion of Microtubule Assembly *in vitro* by Taxol. Nature (London) **277**, 665 (1979).

233. SCHIFF, P.B., and S.B. HORWITZ: Taxol Stabilizes Microtubules in Mouse Fibroblast Cells. Proc. Nat. Acad. Sci. (USA) **77**, 1561 (1980).

234. SCHIFF, P.B., and S.B. HORWITZ: Taxol Assembles Tubulin in the Absence of Exogenous Guanosine-5'-Triphosphate or Microtubule-Associated Proteins. Biochemistry **20**, 3247 (1981).

235. SENILH, V., S. BLECHERT, M. COLIN, D. GUÉNARD, F. PICOT, P. POTIER, and P. VARENNE: Mise en Evidence de Nouveaux Analogues du Taxol Extraits de *Taxus baccata*. J. Nat. Prod. **47**, 131 (1984).

236. SENILH, V., F. GUÉRITTE, D. GUÉNARD, M. COLIN, and P. POTIER: Chimie Organique Biologique. Hemisynthese de Nouveaux Analogues du Taxol. Etude de Leur Interaction Avec la Tubulin. C.R. Acad. Sci. Paris **299**, 1939 (1984).

237. SHEA, K.J., and P.D. DAVIS: The Tricyclo[9.3.1.0^{3,8}]pentadecane System-A Short

Synthesis of a C-Aromatic Taxane Skeleton. Angew. Chem. Int. Ed. Engl. **22**, 419 (1983).

238. SHEA, K.J., and P.D. DAVIS: The Tricyclo[9.3.1.0^{3,8}]pentadecane Ring System. A Short Synthesis of the C-Aromatic Skeleton. Angew. Chem. Suppl. Int. Ed. Engl. **22**, 564 (1983).

239. SHEA, K.J., and J.W. GILMAN: Lewis Acid Catalyzed Intramolecular Diels-Alder Cycloadditions. A Mild Method for the Synthesis of Bicyclo[n.3.1] Bridgehead Alkenes. Tetrahedron Letters **24**, 657 (1983).

240. SHEA, K.J., and J.W. GILMAN: Transition State Conformations of a Lewis Acid Catalyzed Diels-Alder Reaction. The Low-Temperature Cycloaddition of 1-(1-oxo-2-propenyl)-2-(3-isopropenyl-4-methyl-3-pentenyl)benzene. J. Am. Chem. Soc. **107**, 4791 (1985).

241. SHEA, K.J., J.W. GILMAN, C.D. HAFFNER, and T.K. DOUGHERTY: Diasteromeric Atropisomers of the Tricyclo[9.3.1.0^{3,8}]pentadecane Ring System. Synthesis and Structural Studies. J. Am. Chem. Soc. **108**, 4953 (1986).

242. SHEA, K.J., and C.D. HAFFNER: Synthetic Efforts Directed Towards the Taxol Skeleton. The Saturated C-Ring Approach. Tetrahedron Letters **29**, 1367 (1988).

243. SHEA, K.J., R.G. HIGBY, and J.W. GILMAN: Stereochemistry of Hydride Reductions. The Tricyclo[9.3.1.0^{3,8}]pentadecane (Taxane) Ring System. Tetrahedron Letters **31**, 1221 (1990).

244. SHEA, K.J., and S. WISE: Intramolecular Diels-Alder Reactions. A New Entry into Bridgehead Bicyclo[3.n.1]alkenes. J. Am. Chem. Soc. **100**, 6519 (1978).

245. SHIBUYA, H., S. TSUJII, Y. YAMAMOTO, H. MIURA, and I. KITAGAWA: Chemical Transformation of Terpenoids. VII. Synthesis of Chiral Segments, Key Building-Blocks for the Right Half of Taxane-Type Diterpenoids. Chem. Pharm. Bull. **32**, 3417 (1984).

246. SHIBUYA, H., S. TSUJII, Y. YAMAMOTO, K. MURAKAWA, K. TAKAGI, N. KUROKAWA, and I. KITAGAWA: Tennen Yuki Kagobutsu Toronkai Koen Yoshishu, 24th, 340 (1981). (Chem. Abstr. **96**, 218050v (1982)).

247. SIEBURTH, S. McN., and J. CHEN: A Photochemical [4 + 4] Method for the Construction of Annulated Eight-Membered Rings. J. Am. Chem. Soc. **113**, 8163 (1991).

248. SHIRO, M., and H. KOYAMA: Structure of Taxinine: X-ray Analysis of 2,5,9,10-Tetra-O-acetyl-14-bromotaxinol. J. Chem. Soc. 1342 (1971).

249. SHIRO, M., T. SATO, and H. KOYAMA: The Stereochemistry of Taxinine: X-ray Analysis of 2,5,9,10-Tetra-O-acetyl-14-bromotaxinol. Chem. Commun. 97 (1966).

250. SNIDER, B.B., and A.J. ALLENTOFF: Synthesis of the Bicyclo[5.3.1] undecane Moiety (AB Ring System) of Taxanes. J. Org. Chem. **56**, 321 (1991).

251. STASKO, M.W., K.M. WITHERUP, T.J. GHIORZI, T.G. McCLOUD, S. LOOK, G.M. MUSCHIK, and H.J. ISSAQ: Multimodal Thin Layer Chromatographic Separation of Taxol and Related Compounds from *Taxus brevifolia*. J. Liq. Chromatogr. **12**, 2133 (1989).

252. STELLA, V.J., and A.E. MATHEW: Derivatives of Taxol, Pharmaceutical Compositions thereof and Methods for the Preparation thereof. U.S. Patent 4,960,790, October 2, 1990.

253. SUFFNESS, M.: Development of Antitumor Natural Products at the National Cancer Institute. Gann Monograph on Cancer Research **36**, 21 (1989).

254. SUFFNESS, M., and G.A. CORDELL: Taxus Alkaloids. In: The Alkaloids, Chemistry and Pharmacology, Vol. 25, Ed. A. BROSSI, p. 6. New York: Academic Press, 1985.

255. SWINDELL, C.S.: Taxane Diterpene Synthesis Strategies. A Review. Org. Prep. Proced. Int. **23**, 465 (1991).

256. SWINDELL, C.S., and S.F. BRITCHER: Construction of the Taxane C-Ring Epoxy Alcohol Moiety and Examination of its Possible Involvement in the Biogenesis of the Taxane 3-Oxetanol Structure. J. Org. Chem. 51, 793 (1986).

257. SWINDELL, C.S., T.F. ISAACS, and K.J. KANES: Bicyclo[5.3.1]Undec-1(10)-ene Bridgehead Olefin Stability and the Taxane Bridehead Olefin. Tetrahedron Letters 26, 289 (1985).

258. SWINDELL, C.S., and S.J. deSOLMS: Synthesis of the Taxane Diterpenes: Construction of a BC Ring Intermediate for Taxane Synthesis. Tetrahedron Letters 25, 3801 (1984).

259. SWINDELL, C.S., and S.J. deSOLMS: Regio- and Stereochemistry of the Intramolecular [2 + 2] Photoproducts from Two Related Vinylogous Imides. Tetrahedron Letters 25, 3797 (1984).

260. SWINDELL, C.S., and B.P. PATEL: Synthesis and Stereochemistry of a Saturated Tricyclic Taxane Model. Tetrahedron Letters 28, 5275 (1987).

261. SWINDELL, C.S., and B.P. PATEL: Stereoselective Construction of the Taxinine AB System through a Novel Tandem Aldol-Payne Rearrangement Annulation. J. Org. Chem. 55, 3 (1990).

262. SWINDELL, C.S., B.P. PATEL, J.S. deSOLMS, and J.P. SPRINGER: A Route for the Construction of the Taxane BC Substructure. J. Org. Chem. 52, 2346 (1987).

263. SWINDELL, C.S., N.E. KRAUSS, S.B. HORWITZ, and I. RINGEL: Biologically Active Taxol Analogues with Deleted A-Ring Side Chain Substituents and Variable C-2' Configurations. J. Med. Chem. 34, 1176 (1991).

264. TAGA, J.: Über Taxinin and Anhydrotaxininol. Chem. Pharm. Bull. 8, 934 (1960).

265. TAGA, J.: Über die Konstitution des Anhydrotaxinols. Chem. Pharm. Bull. 12, 389 (1964)

266. TAKAHASHI, T.: Japanese Yew Leaves. III. Taxinin I. Yakugaku Zasshi 51, 401 (1931).

267. TAKAHASHI, T., K. UEDA, R. OISHI, and K. MINAMOTO: Über Taxinin. Chem. Pharm. Bull. 6, 728 (1958).

268. TAKOUDJU, M., M. WRIGHT, J. CHENU, F. GUÉRITTE-VOEGELEIN, and D. GUÉNARD: Absence of 7-Acetyl Taxol Binding to Unassembled Brain Tubulin. FEBS Lett. 227, 96 (1988).

269. TAYLOR, D.A.: An Extractive from Taxus baccata. West Afr. J. Biol. Appl. Chem. 7, 1 (1963).

270. TEKOL, Y.: Negative Chronotropic and Atrioventricular Blocking Effects of Taxine on Isolated Frog Heart and its Acute Toxicity in Mice. Planta Med. 357 (1985).

271. TEKOL, Y.: Acute Toxicity of Taxine in Mice and Rats. Vet. Hum. Toxicol. 33, 337 (1991).

272. TEKOL, Y.: A. ERENEMEMISOGLU, and K. SUNGUROGLU: Protective Effect of Taxine Against Isoproterenol Induced Myocardial Damage in Rats. Acta Pharm. Turcica 33, 67 (1991).

273. TEKOL, Y., and M. KAMEYAMA: Elektrophysiologische Untersuchungen über den Wirkungsmechanismus des Eibentoxins Taxin auf das Herz. Arzneim.-Forsch. 37, 428 (1987).

274. THIGPEN, T., J. BLESSING, H. BALL, S. HUMMEL, and R. BARRET: Phase II Trial of Taxol as Second-line Therapy for Ovarian Carcinoma: a Gynecologic Oncology Group Study. Proc. Amer. Soc. Clin. Oncol. 9, 604 (1990).

275. TROST, B.M., and M.J. FRAY: Studies Directed Toward Taxanes. Preparation of α-Ketols by Oxidative Ring-Opening of Epoxides. Tetrahedron Letters 29, 2163 (1988)

276. TROST, B., and H. HEIMSTRA: Ion Pair Effects in an Intercalation Process. An Approach to the Bicyclo[5.3.1]undecyl System of Taxane. J. Am. Chem. Soc. 104, 886 (1982).

277. TROST, B., and H. HEIMSTRA: Synthetic Studies Toward Taxanes. A Sulfur Based Approach to the Benzilic Acid Rearrangement. Tetrahedron **42**, 3323 (1986).
278. TROST, B.M., and M.J. FRAY: A Three Carbon Intercalation of an Enediol Silyl Ether; A Short Entry to the Bicyclo[5.3.1]undecyl System of Taxanes. Tetrahedron Letters **25**, 4605 (1984).
279. UEDA, K., S. UYEO, Y. YAMAMOTO, and Y. MAKI: The Structure of Taxinine, A Nitrogen-Free Compound Occurring in *Taxus cuspidata*. Tetrahedron Letters, 2167 (1963).
280. UYEO, S., Y. MAKI, and Y. YAMAMOTO: Taxine XIV. The Structure of Isopropyliden-edihydrotaxinolactone, an Autooxidation Product of Isopropylidenedihydrotaxinol. Chem. Pharm. Bull. **14**, 502 (1966).
281. UYEO, S., K. UEDA, Y. YAMAMOTO, N. HAZAMA, and Y. MAKI: Taxine VIII. Taxinine and Taxinol. Yakugaku Zasshi **82**, 1081 (1962).
282. YEH, M.-K., J.-S. WANG, L-P. LIU, and F.-C. CHEN: Some Taxane Derivatives from the Heartwood of *Taxus mairei*. J. Chin. Chem. Soc. **35**, 309 (1988).
283. VIDENSEK, N., P. LIM, A. CAMPBELL, and C. CARLSON: Taxol Content in Bark, Wood, Root, Leaf, Twig, and Seedling from Several *Taxus Species*. J. Nat. Prod. **53**, 1609 (1990).
284. VOHORA, S. B., and I. KUMAR. Studies on *Taxus baccata*: 1. Preliminary Phytochemi-cal and Behavioral Investigations. Planta Med. **20**, 100 (1971).
285. WANI, M.C., H.L. TAYLOR, M.E. WALL, P. COGGON, and A.T. MCPHAIL: Plant Antitumor Agents VI. The Isolation and Structure of Taxol, a Novel Antileukemic and Antitumor Agent from *Taxus brevifolia*. J. Am. Chem. Soc. **93**, 2325 (1971).
286. WEISS, R.B., R.C. DONEHOWER, P.H. WIERNIK, T. OHNUMA, R.J. GRALLA, D.L. TRUMP, J.R. BAKER, Jr., D.A. VAN ECHO, D.D. VON HOFF, and B. LEYLAND-JONES: Hypersensitivity Reactions from Taxol. J. Clin. Oncol. **8**, 1263 (1990).
287. WENDER, P.A., and N.C. IHLE: Nickel-Catalyzed Intramolecular [4 + 4] Cycloaddi-tions: A New Method for the Synthesis of Polycycles Containing Eight-Membered Rings. J. Am. Chem. Soc. **108**, 4678 (1986).
288. WENDER, P.A., and N.C. IHLE: Nickel-Catalyzed Intramolecular [4 + 4] Cycloaddi-tions: 2. Allylic Stereoindiction and Modeling Studies in the Preparation of Bicyclo-[6.4.0]dodecadienes. Tetrahedron Letters **28**, 2451 (1987).
289. WENDER, P.A., and M.J. TEBBE: Nickel(O)-Catalyzed Intramolecular [4 + 4] Cyc-loadditions: 5. The Type II Reaction in the Synthesis of Bicyclo[5.3.1]undecadienes. Synthesis, 1089 (1991).
290 WENDER, P.A., and M.L. SNAPPER: Intramolecular Nickel Catalyzed Cycloadditions of *Bis*-dienes: 3 Approaches to the Taxane Skeleton. Tetrahedron Letters **28**, 2221 (1987).
291. WHEELER, N.C., K. JECH, S. MASTERS, S.W. BROBST, A.B. ALVARADO, A.J. HOOVER, and K.M. SNADER: Effects of Genetic, Epigenetic, and Environmental Factors on Taxol Content in *Taxus brevifolia* and Related Species. J. Nat. Prod. **55**, 432 (1992).
292. WIERNIK, P.H., E.L. SCHWARTZ, A. ENZIG, J.J. STRAUMAN, R.B. LIPTON, and J.P. DUTCHER: Phase I Trial of Taxol Given as a 24-Hour Infusion Every 21 Days: Responses Observed in Metastatic Melanoma. J. Clin. Oncol. **5**, 1232 (1987).
293. WIERNIK, P.H., E.L. SCHWARTZ, J.J. STRAUMAN, J.P. DUTCHER, R.B. LIPTON, and E. PAIETTA: Phase I Clinical and Pharmacokinetic Study of Taxol. Cancer Res. **47**, 2486 (1987).
294. WINKLER, J.D., and J.P. HEY: Inside-Outside Stereoisomerism: The Synthesis of *trans*-Bicyclo[5.3.1]undecane-11-one. J. Am. Chem. Soc. **108**, 6425 (1986).
295. WINKLER, J.D., J.P. HEY, and S.D. DARLING: Studies Directed Towards the Synthesis

of Taxane Diterpenes: A remarkable Rearrangement. Tetrahedron Letters **27**, 5959 (1986).

296. WINKLER, J.D., C. LEE, L. RUBO and C.L. MULLER: Stereoselective Synthesis of the Tricyclic Skeleton of the Taxane Diterpenes. The First C-Silylation of a Ketone Enolate. J. Org. Chem. **54**, 4491 (1989).

297. WINKLER, J.D., B. HONG, J.P. HEY, and P.G. WILLIARD: Inside-Outside Stereoisomerism. 5. Synthesis and Reactivity of *trans*-bicyclo[n.3.1] alkanes Prepared via the Intramolecular Photocycloaddition of Dioxenones. J. Am. Chem. Soc. **113**, 8839 (1991).

298. WINTERSTEIN, E. and A. GUYER: Weitere Beitrage zur Kenntnis des Taxins. Z. Physiol. Chem. **128**, 175 (1923).

299. WINTERSTEIN, E. and D. IATRIDES: Uber das aus *Taxus baccata*, Eibe, Darstellbare Alkaloid, Taxin. Z. Physiol. Chem. **117**, 240 (1921).

300. WITHERUP, K.M., S.A. LOOK, M.W. STASKO, T.G. McCLOUD, H.J. ISSAQ, and G.M. MUSCHIK: High Performance Liquid Chromatographic Separation of Taxol and Related Compounds from *Taxus brevifolia*. J. Liq. Chromatogr. **12**, 2117 (1989).

301. WITHERUP, K.M., S.A. LOOK, M.W. STASKO, T.J. GHIORZI, G.M. MUSCHIK, and G.M. CRAGG: Taxus spp. Needles Contain Amounts of Taxol Comparable to the Bark of *Taxus brevifolia*: Analysis and Isolation. J. Nat. Prod. **53**, 1249 (1990).

302. WOODS, M.C., H.-C. CHIANG, Y. NAKADAIRA, and K. NAKANISHI: The Nuclear Overhauser Effect, a Unique Method of Defining the Relative Stereochemistry and Conformation of Taxane Derivatives. J. Am. Chem. Soc. **90**, 522 (1968).

303. XU, L.X. and A.R. LIU: Determination of Taxol in the Extract of *Taxus chinesis* by reversed phase HPLC. Acta Pharm. Sin. (Yaoxue Xuebao) **24**, 552 (1989).

304. YADAV, J.S., and R. RAVISHANKAR: A Novel Approach Towards the Synthesis of Functionalized Taxane Skeleton Employing Wittig Rearrangement. Tetrahedron Letters **32**, 2629 (1991).

305. YAMAMOTO, Y., S. UYEO, and K. UEDA: Structure of Anhydrotaxininol. Chem. Pharm. Bull. **3**, 386 (1964).

306. YEH, M.-K., J.-S. WANG, L.-P.LIU, and F.-C. CHEN: A New Taxane Derivative from the Heartwood of *Taxus mairei*. Phytochem. **27**, 1534 (1988).

307. YEH, M.-K., J.-S. WANG, L.-P. LIU, and F.-C. CHEN: Some Taxane Derivatives from the Heartwood of *Taxus mairei*. J. Chin. Chem. Soc. **35**, 309 (1988).

308. YEH, M.-K., J.-S. WANG, W.L. YANG, and F.-C. CHEN: A New Taxane Derivative from the Heartwood of *Taxus mairei*. Proc. Natl. Sci. Counc., Rep. China, Part A: Phys. Sci. Eng. **12**, 89 (1988).

309. YOSHIZAKI, F., M. FUKUDA, S. HISAMICHI, T. ISHIDA, and Y. IN: Structures of Taxane Diterpenoids from the Seeds of Japanese Yew, *Taxus cuspidata*. Chem. Pharm. Bull. **36**, 2098 (1988).

310. ZHANG, Z, and Z. JIA: Taxanes from *Taxus yunnanensis*. Phytochem. **29**, 3673 (1990).

311. ZHANG, Z., and Z. JIA: Taxanes from *Taxus chinensis*. Phytochem. **30**, 2345 (1991).

312. ZHANG, Z.P., and Z.J. JIA: New Taxanes From *Taxus chinensis*. Z. Chin. Chem. Lett. **1**, 91 (1990).

313. ZHANG, Z.-P., Z.-J. JIA, Z.-Q. ZHU, Y.-X. CUI, J.-L. CHENG, and Q.-G. WANG: Studies on the Chemical Constituents of *Taxus*. Chinese Sci. Bull. **21**, 1630 (1989).

314. ZHANG, Z., Z. JIA, Z. ZHU, Y.CUI, J. CHENG, and Q. WANG: New Taxanes from *Taxus chinesis*. Planta Med. **56**, 293 (1990).

315. ZHAO, Z., D.G.I. KINGSTON, and A.R. CROSSWELL: Modified Taxols, 6. Preparation of Water-Soluble Prodrugs of Taxol. J. Nat. Prod. **54**, 1607 (1991).

316. ZUCKER, P.A., and J.A. LUPIA: An Oxy-Cope Route to the Taxane AB Ring System. Synlett 729 (1990).

Addendum references

S1. APPENDINO, G., P. GARIBOLDI, B. GABETTA, R. PACE, E. BOMBARDELLI, and D. VITERBO. 14-β-Hydroxy-10-deacetylbaccatin III, a New Taxane from Himalayan Yew (*Taxus wallichiana Zucc.*). Chem. Soc., Perkin Trans. 1, in press (1992).

S2. APPENDINO, G., P. GARIBOLDI, A. PISETTA, E. BOMBARDELLI, and B. GABETTA. Taxanes from *Taxus baccata*. Phytochemistry, in press (1992).

S3. APPENDINO, G., P. LUZZO, P. GARIBOLDI, E. BOMBARDELLI, and B. GABETTA. A3,11-Cyclotaxane from *Taxus baccata*. Phytochemistry, in press (1992).

S4. APPENDINO, G., S. TAGILIAPIETRA, H.C. OZEN, P. GARIBOLDI, B. GABETTA, and E. BOMBARDELLI. Taxanes from the Seeds of *Taxus Baccata L.* J. Nat. Prod. **55**, in press (1993).

S5. AURIOLA, S.O.K., A.-M. LEPIST, T. NAARANLAHTI, and S.P. LAPINJOKI: Determination of Taxol by High-performance Liquid Chromatography-Thermospray Mass Spectrometry. J. Chromatogr. **594**, 153 (1992).

S6. BAKER, J.K.: Nuclear Overhauser Effect Spectroscopy (NOESY) and Dihedral Angle Measurements in the Determination of the Conformation of Taxol in solution. Spectroscopy Letters **25**, 31 (1992).

S7. BENCHIKH-LE-HOCINE, M., D. DO KHAC, and M. FÉTIZON: Model Studies in Taxane Diterpene Synthesis. Part III. Synth. Comm. **22**, 245 (1992).

S8. BENCHIKH-LE-HOCINE, M., D. DO KHAC, M. FÉTIZON, and T. PRANGE: A Photochemical Approach to a Potential Precursor of the Bicyclo[6.4.0]dodecane Ring System of Taxane Diterpenes. Synth. Commun. **22**, 1871 (1992).

S9. BLECHERT, S., R. MÜLLER, and M. BEITZEL: Stereoselective Synthesis of Taxol Derivative. Tetrahedron **48**, 6953 (1992).

S10. CHATTERJEE, A.,J.S. WILLIAMSON, J.K. ZJAWIONY, and J.R. PETERSON: Synthesis of a Photoreactive Taxol Side Chain. Bioorg. Med. Chem. Letters **2**, 91 (1992).

S11. CHEN, W.M., P.L. ZHANG, B.WU, and Q.T. ZHANG: Studies on the Chemical Constituents of *Taxus yunnanensis*. Acta Pharm. Sin. **26**, 747 (1991).

S12. CHEN, W.M., P.L. ZHANG, B.WU, and Q.T. ZHANG: Studies on the Chemical constituents of *Taxus yunnanensis*. Chin. Chem. Lett. **2**, 441 (1991).

S13. CHUANG, L.C., K.J. CHEN, Y.S. LIN, and F.C. CHEN: Reinvestigation on the Constituents of the Heartwood of Taiwan Yew. T'ai-wan K'o Hsueh **42**, 29 (1989). Chem. Abstr. **112**, 175629v (1990).

S14. CHUANG, L.C., K.J.CHEN, Y.S.LIN, and F.C. CHEN: Constituents of the Heartwood of the Taiwan Yew. Part IV. Isolation of 1,4-*p*-Menthanediol and 1-Dehydroxybaccatin IV. Huaxue, **48**, 275 (1990). Chem. Abstr. **116**, 148178y (1992).

S15. COMMERCON, A., D. BEZARD, F. BERNARD, and J.D. BOURZAT: Improved Protection and Esterification of a Precursor of the Taxotere and Taxol Side Chains. Tetrahedron Letters, **33**, 5185 (1992).

S16. DENG, L., and E.N. JACOBSEN: A Practical, Highly Enantioselective Synthesis of the Taxol Side Chain via Asymmetric Catalysis. J. Org. Chem. **57**, 4320 (1992).

S17. FARINA, V., S.I. HAUCK, and D.G. WALKER: A Simple Chiral Synthesis of the Taxol Side Chain. Synlett, 761 (1992).

S18. FARINA, V., and S. HUANG: The Chemistry of Taxanes: Unexpected Rearrangement of Baccatin III During Chemoselective Debenzoylation with $Bu_3SnOMe/LiCl$. Tetrahedron Letters **33**, 3979 (1992).

S19. FUNK, R.L., J.A. WOS, and W.J. DAILY: A Successive Ring Annelation Strategy for Construction of the Taxane Ring System. 204th National Meeting of the American Chemical Society, Division of Organic Chemistry, (Washington, DC, 1992). Abstract 255.

S20. FURUKAWA, T., K. MORIHIRA, Y. HORIGUCHI, and I. KUWAJIMA: Synthetic Studies on Taxane Carbon Framework. A Highly Efficient Eight-membered Ring Cyclization with Complete Stereocontrol. Tetrahedron **34**, 6975 (1992).

S21. GEORG, G.I., Z.S. CHERUVALLATH, R.H. HIMES, and M.R. MEJILLANO: Novel Biologically Active Taxol Analogues: Baccatin III 13-(N-(p-Chlorobenzoyl)-2'R,3'S)-3'-phenylisoserinate) and Baccatin III 13-(N-Benzoyl-(2'R,3'S)-3'-(p-chlorophenyl)-isoserinate). Bioorg. Med. Chem. Letters **2**, 295 (1992).

S22. GEORG, G.I., Z.S. CHERUVALLATH, R.H. HIMES, and M.R. MEJILLANO and C.T. BURKE: Synthesis of Biologically Active Taxol Analogues with Modified Phenylisoserine Side Chains. J. Med. Chem. **35**, 4230 (1992).

S23. GUNAWARDANA, G.P., U. PREMACHANDRAN, N.S. BURRES, D.N. WHITTERN, R. HENRY, S. SPANTON, and J.B. McALPINE. Isolation of 9-Dihydro-13-acetylbaccatin III from *Taxus canadensis*. J. Nat. Prod. **55**, in press (1992).

S24. HARVEY, S.D., J.A. CAMPBELL, R.G. KELSEY, and N.C. VANCE: Separation of Taxol from Related Taxanes in *Taxus brevifolia* Extracts by Isocratic Elution Reversed-phase Microcolumn High-performance Liquid Chromatography. J. Chromatogr. **587**, 300 (1991).

S25. HITCHCOCK, S.A., and G. PATTENDEN: A Tandem Radical Macrocyclisation-Radical Transannulation Strategy to the Taxane Ring System. Tetrahedron Letters **33**, 4843 (1992).

S26. HORWITZ, S.B. Mechanism of Action of Taxol. Trends Pharm. Sci., **13**, 134 (1992).

S27. JACKSON, R.W., R.G. HIGBY, J.W. GILMAN, and K.J. SHEA: The Chemistry of C-aromatic Taxane Derivatives: Atropisomer Control of Reaction Stereochemistry. Tetrahedron **48**, 7013 (1992).

S28. JIA, Z.-J., and Z.-P. ZHANG: Taxanes from *Taxus chinensis* III. Chin. Sci. Bull. **36**, 1967 (1991).

S29. KELSEY, R.G., and N.C. VANCE: Taxol and Cephalomannine Concentrations in the Foliage and Bark of Shade-grown and Sun-exposed *Taxus brevifolia* Trees. J. Nat. Prod. **55**, 912 (1992).

S30. LANSING, A., M. HAERTEL,. M. GORDON, and H.G. FLOSS: Biosynthetic Studies on Taxol. Planta Med. **57**, Supplement 2, A83 (1991).

S31. LIANG, J.,and D.G.I. KINGSTON: Two New Taxane Diterpenoids from *Taxus mairei*. J. Nat. Prod., in press (1993).

S32. MAGEE, T.V., W.G. BORNMANN, R.C.A.. ISAACS, and S. DANISHEFSKY: A straightforward Route to Functionalized Intermediates Containing the CD Structure of Taxol. J. Org. Chem. **57**, 3274 (1992).

S33. McCLURE, T.D., K.H. SCHRAM, and M.L.J. REIMER: The Mass Spectrometry of Taxol. J. Am. Soc. Mass Spectrom. **3**, 672 (1992).

S34. MEHTA, G., R. BARONE, P. AZARIO, F. BARBERIS, M. ARBELOT, and M. CHANON: New Computer-Based Approach for Seeking a Key Step in the Synthesis of Complex Structures. Application to Taxane and Crinipellin Diterpenoid Frameworks. Tetrahedron **41**, 8953 (1992).

S35. NICOLAOU, K.C., C.-K. HWANG, E.J. SORENSEN, and C.F. CLAIRBORNE: A Convergent Strategy Towards Taxol. A Facile Enantioselective Entry Into a Fully Functionalized Ring A System. J. Chem. Soc., Chem. Commun., 1117 (1992).

S36. NICOLAOU, K.C., J.J. LIU, C.-K. HWANG, W.-M. DAI, and R.K. GUY: Synthesis of a Fully Functionalized CD Ring System of Taxol. J. Chem. Soc., Chem. Commun., 1118 (1992).

S37. OJIMA, I., I. HABUS, M. ZHAO, M. ZUCCO, Y.H. PARK, C.M. SUN, and T. BRIGAUD: New and Efficient Approaches to the Semisynthesis of Taxol and Its C-13 Side Chain Analogs by Means of β-lactam Synthon Method. Tetrahedron **48**, 6985 (1992).

S38. PAQUETTE, L.A., M. ZHAO, and D.FRIEDRICH: Unprecedented Intramolecular Redox Behavior of a Taxane Derivative. Tetrahedron Letters, in press.

S39. PAQUETTE, L.A., K.D. COMBRINK, S.W. ELMORE, and M. ZHAO: Setting the Bridge-head Oxidation Level in *trans*-Tricyclo[9.3.1.0$^{3.8}$]pentadecanes as a Prelude to the Dual Synthesis of Taxol and Taxusin. Helv. Chim. Acta, in press.

S40. PAQUETTE, L.A., S.W. ELMORE, K.D. COMBRINK, and E.R. HICKEY: An Enantiospecific Approach to the Taxanes, Direct Access to Functionalized *cis*-Tricyclo[9.3.1.0$^{3.8}$]pentadecanes via α-Ketol and Wagner-Meerwein Rearrangements. Helv. Chim. Acta, in press.

S41. POTIER, P.: Search and Discovery of New Antitumour Compounds. J. Chem. Soc., Perkin Trans. 1, 113 (1992).

S42. QUENEAO, Y., W.J. KROL, W.G. BORNMANN, and S.J. DANISHEFSKY: A Ready Synthesis of Intermediates Containing the A-Ring Substructure of Taxol: A Diels-Alder Route to the B-seco Taxane Series. J. Org. Chem. **57**, 4043 (1992).

S43. SAHA, G., A. KARPHA, S.S. ROY and S. GHOSH: A Simple Approach to the Bicyclo[5.3.1]undecane System Present in Taxanes. J. Chem. Soc. Perkin Tans. 1, 1587 (1992).

S44. SAMARANAYAKE, G., K. NEIDIGH, and D.G.I. KINGSTON: Modified Taxols, 8. Deacylation and Reacylation of Baccatin III. J. Nat. Prod., submitted for publication.

S45. SHEA, K.J., and S.T. SAKATA: Reversal of Regiospecificity in the Kinetic vs. Thermodynamic Enolization of Bicyclic Ketones. Direct Bridgehead Functionalization of the Bicyclo[5.3.1]Undecane Ring System. Tetrahedron Letters. **33**, 4261 (1992).

S46. SLICHENMYER, W.J., and D.D. VON HOFF: Anti-Cancer Drugs **2**, 519 (1991).

S47. VANHAELEN-FASTRE, R., B. DIALLO, M. JAZIRI, M.L. FAES, J. HOMES, and M. VANHAELEN: High-speed Countercurrent Chromatography Separation of Taxol and Related Diterpenoids from *Taxus baccata.* J. Liq. Chromatogr. **15**, 697 (1992).

S48. WAHL, A., F. GUERITTE-VOEGELEIN, D. GU NARD, M.-T. LE GOFF, and P. POTIER. Rearrangement Reactions of Taxanes: Structural Modifications of 10-Deacetylbaccatin III. Tetrahedron **48**, 6965 (1992).

S49. WENDER, P.A., and T.P. MUCCIARO: A New and Practical Approach to the Synthesis of Taxol and Taxol Analogues: The Pinene Path. J. Am. Chem. Soc. **114**, 5878 (1992).

S50. WENDER, P.A., and D.B. RAWLINS: Toward the Synthesis of the Taxol C, D Ring System: Photolysis of α-Methoxy Ketones. Tetrahedron **48**, 7033 (1992).

S51. WINKLER, J.D., and D. SUBRAHMANYAM: Studies Directed Toward the Synthesis of Taxol: Preparation of C-13 Oxygenated Taxane Congeners. Tetrahedron **48**, 7049 (1992).

S52. XU, L.X., and A.R. LIU: Determination of Taxol in *Taxus chinensis* by HPLC Method. Acta Pharm. Sin. (Yaoxue Xuebao) **26**, 537 (1991).

S53. ZAMIR, L.O., M.E. NEDEA, S. BELAIR, F. SAURIOL, Ö. MAMER, E. JACQMAIN, F.I. JEAN, and F.X. GARNEAU: Taxanes Isolated from *Taxus canadensis.* Tetrahedron Letters **33**, 5173, 6548 (1992).

S54. ZAMIR, L.O., M.E. NEDA, and F.X. GARNEAU: Biosynthetic Building Blocks of *Taxus canadensis* Taxanes. Tetrahedron Letters **33**, 5235 (1992).

S55. ZHANG, Z., and Z. JIA: New Taxane Diterpenes from *Taxus chinensis. Huaxue Xuebao*, **49**, 1023 (1991). Chem. Abstr. **116**, 148188b (1992).

(Received May 4, 1992)

Author Index

Page numbers printed in *italics* refer to References

Adeline, M.-T. *182*
Ahond, A. *177*
Aizpurua, J.M. *184*
Allentoff, A.J. 140, *186*
Alvarado, A.B. *188*
Amarasekara, A.S. *174*
Andriamialisoa, R.Z. *173*
Appendino, G. 166, *190*
Arbelot, M. *191*
Armstrong, D.K. *182*
Arnold, E. *183*
Arrieta, A. *184*
Auriola, S.O.K. *190*
Aurrekoetxea, N. *184*
Azario, P. *191*

Babirad, S.A. *185*
Baker, J.K. *190*
Baker Jr., J.R. *188*
Ball, H. *187*
Balza, F. *173*
Barberis, F. *191*
Barman, P. *184*
Barone, R. *191*
Barret, R. *187*
Barrios, H. *173*
Bauereis, R. *173*
Baxter, J.N. *173*
Begley, M.J. *174*
Beitzel, M. *190*
Belair, S. *191*
Beloeil, J. *175*
Benchikh-le-Hocine, M. *174, 190*
Benesi, A.J. *177*
Benjamin, R.S. *181*
Berkowitz, W.F. *127–130*, 151, *174*

Bernard, F. *190*
Bertholdt, H. *178*
Beutler, J.A. 39, 173, *174, 175*
Bezard, D. *190*
Bhattacharya, A. *185*
Bissery, M.-C. *174*
Bjamer, K. *175*
Blechert, S. 7, 100, 125, 150, 171, *174, 180, 183, 185, 190*
Blessing, J. *187*
Blount, J.F. *175*
Blume, E. *174*
Bodem, G.B. 154, *181*
Bombardelli, E. *190*
Bonnert, R.V. *174*
Bornmann, W.G. *191, 192*
Bourzat, J.D. *190*
Breitmaier, E. *178*
Brigand, T. *191*
Britcher, S.F. 157, *187*
Brobst, S.W. *175, 179, 188*
Brown, P.A. *174*
Brown, T. *184*
Brundrett, R.B. *182*
Bryan-Brown, T. *174*
Büchi, G. 96, *174*
Burke, B.A. *175*
Burke, C.T. *191*
Burke, P.J. *185*
Burres, N.S. *191*
Buzdar, A.U. *179*

Caesar, J. *175*
Callow, R.K. *175*
Campbell, A. *188*
Campbell, J.A. *191*

Cardellina II, J.H. *175*
Carlson, C. *188*
Cassady, R.E. *185*
Castellano, E.E. *175, 176*
Cazenave, L.A. *185*
Cervantes, H. *174, 175*
Cesario, M. *178*
Chan, W.R. *175*
Chang, C.-J. *179*
Chanon, M. *191*
Chatterjee, A. *190*
Chaudhury, R.R. *175*
Chauviere, G. *175*
Chen, F.-C. *179, 182, 188–190*
Chen, J. 110, *186*
Chen, K.J. *190*
Chen, T.L. *182*
Chen, W.M. 7, *175, 184, 190*
Cheng, J.-L. *189*
Chenu, J. *175, 187*
Cheruvallath, Z.S. *191*
Chiang, H.-C. *175, 189*
Chmurny, G.N. 173, *174, 175*
Christian, M.C. *185*
Chuang, L.C. *190*
Clairborne, C.F. *191*
Clardy, J. *183*
Clark, G.R. 126, 127, 152, *175, 181*
Coffey, V.A. *183*
Coggon, P. *188*
Colin, M. *175, 185*
Combrink, K.D. *177, 192*
Commercon, A. 172, *190*
Convert, O. *177*
Cook, P.M. *185*
Cooks, R.G. 34, 53, *179*
Cordell, G.A. 7, *186*
Correa, A. *176*
Cossio, F.P. *184*
Cragg, G.M. *189*
Craven, B.M. *185*
Cros, S. *183*
Crosswell, A.R. *189*
Cui, Y.-X. *189*

Dai, W.-M. *191*
Daily, W.J. *177, 190*
Danishefsky, S.J. 170, *191, 192*
Darling, S.D. *188*
Dauben, W.G. *175*

David, B. *178*
Davis, P.D. *185, 186*
DeAmicis, C.V. 150, *175*
De Brabander, M. *176*
Del Campo, A.A. *178*
Della Casa De Marcano, D.P. *176, 184*
De Mendez, J. *176, 184*
De Mey, J. *176*
Deng, L. 172, *190*
Denis, J.-N. *176*
DeRussy, D.T. *184*
DeSolms, S.J. *187*
Deutsch, H.M. *176, 179*
Diallo, B.M. *179, 192*
Dicosmo, F. *183*
Do, H.D. *184*
Do Khac, D. *174, 175, 190*
Donehower, R.C. *176, 182, 185, 188*
Dorn, H.C. *176*
Dougherty, T.K. *186*
Duchesne, J.P. *176*
Dudani, A.K. *178*
Dukes, M. *177*
Dutcher, J.P. *177, 188*

Einzig, A.I. *177, 188*
Elmore, S.W. *177, 192*
Erdtman, H. *177*
Erenememisoglu, A. *187*
Ettinger, D.S. *176, 182, 184*
Ettouati, L. *177*
Eyre, D.H. *177*
Eyre, E.D. *177*

Faes, M.L. *192*
Falzone, J.J. *177*
Fang, Q.C. *184*
Fant, J. *185*
Farina, V. 172, *190*
Fawcett, J. *174*
Ferguson, G. *175*
Fétizon, M. 99, 100, 103, 104, 116–118, 133,
    135, 168, 172, *173–175, 177, 190*
Fillion, H. 150, *177*
Fisherman, J.S. *185*
Floss, H.G. *191*
Forman, A.D. *179*
Frahm, A.W. *180*
Fray, M.J. *187, 188*
Frecknall, E.A. *174*

Frejd, T. 85, 86, *177, 184*
Friedrich, D. *192*
Fronczek, F.R. *185*
Frye, D.K. *179*
Fujioka, H. *180, 181*
Fukaya, C. *179*
Fukuda, M. *189*
Funk, R.L. 120, 121, 172, *177, 190*
Furukawa, T. *177, 179, 191*

Gabetta, B. *190*
Gadwood, R.C. 104, *177*
Garcia-Arenas, R. *182*
Gares, M. *183*
Garg, S.K. *175*
Gariboldi, P. *190*
Garl, S. *177*
Garneau, F.X. *192*
Gates, A.E. *182*
Gecu, A.G. *179*
Georg, G.I. *184, 191*
Geuens, G. *176*
Ghiorzi, T.J. *186, 189*
Ghosh, S. 134, 136, 172, *185, 192*
Gilman, J.W. *186, 191*
Glinski, J.A. *176*
Goldberg, G. *177*
Gordon, M. *191*
Gorowski, E. *177*
Graf, E. 5, *178*
Gralla, R.J. *181, 188*
Greene, A.E. 81, 145, 147, 148, 150, *176*
Greenspan, E.M. *183*
Grem, J.L. *178*
Grochow, L.B. *176, 182*
Grumbine, F.C. *182*
Gu, J. *178*
Guarineri, T. *185*
Guénard, D. 7, *174–176, 178, 182, 183, 185, 187*
Guéritte, F. *185*
Guéritte-Voegelein, F. 7, *174–176, 178, 182, 183, 187, 192*
Guilhem, J. *178*
Guir, F. *174, 175*
Gulland, J.M. *175*
Gunatilaka, A.A.L. *180*
Gu Nard, D. *191*
Gunawardana, G.P. *191*
Guo, Y. 174
Gupta, R.S. *178*

Guy, R.K. *191*
Guyer, A. *189*

Habus, I. *184, 191*
Haertel, M. *191*
Haffner, C.D. 89, *186*
Halsall, T.G. 5, 79, 157, *175, 176*
Hamel, E. *178*
Hanna, I. *173, 174, 177*
Harada, K. *178*
Harrison, J.W. *177, 178*
Hartzell Jr., H. *178*
Harusawa, S. *179*
Harvey, S.D. *191*
Haslam, E. 154, *184*
Hatanaka, Y. *178*
Haugwitz, R.D. *176*
Hauck, S.I. *190*
Hawkins, D.R. *180*
Hawkinson, S.W. *185*
Hayakawa, K. *178*
Hazama, N. *188*
Heimstra, H. 121, *187, 188*
Heinstein, P. *179*
Henry, R. *191*
Henry, T.A. *178*
Hernandez, M. *176*
Hey, J.P. *188, 189*
Hickey, E.R. *192*
Higby, R.G. *186, 191*
Hilton, B.D. *175*
Himes, R.H. *182, 191*
Hirose, Y. *181*
Hisamichi, S. *189*
Hitchcock, S.A. 171, *191*
Ho, T.-I. *179*
Hodder, O.J.R. *175, 176*
Hodges, J.C. *180*
Hoke III, S.H. *179*
Holland, J.F. *183*
Holmes, F.A. *179*
Holton, R.A. 82, 94–96, 150, 170, *179*
Homes, J. *179, 192*
Hon, Y. *181*
Hong, B. *189*
Hönig, H. 146, *179*
Honkan, V.A. *175*
Hoover, A.J. *188*
Hori, M. *180, 181*
Horiguchi, Y. *177, 179, 191*

Hornby, G.M. *175, 176*
Hortobagyi, G.N. *179*
Horwitz, S.B. 6, 161, 164, *179, 182, 184, 185,*
    187, 191
Houk, K.N. *185*
Huang, C.H.O. *179*
Huang, S. *190*
Hubbell, J.P. *175*
Hummel, S. *187*
Huttel, M.S. *179*
Hwang, C.-K. *191*

Iatrides, D. *189*
Igarashi, A. *180*
Ihle, N.C. *188*
Iinuma, M. *181*
Inouye, Y. 123–125, 129–131, *179, 181*
Isaacs, R.C.A. *191*
Isaacs, T.F. *187*
Ise, F. *181*
Ishida, T. *189*
Issaq, H.J. *186, 189*
Ivey, C.A. *180*

Jackson, C.B. *174, 179*
Jackson, R.W. *191*
Jacobsen, E.N. 172, *190*
Jacqmain, E. *192*
Jansen, M. *183*
Jayasinghe, L.R. *184*
Jayasuria, H. *179*
Jaziri, M. *179, 192*
Jean, F.I. *192*
Jech, K. *188*
Jenkins, P.R. 90–92, *174*
Jennings, D.W. *179*
Jia, Z.-J. *178, 180, 189, 191, 192*
Jiang, H. *183*
Jitrangsri, C. *180, 182, 185*
Johnson, S. *180*
Jungheim, L.N. *180*
Juo, R.R. *179*

Kaczmarek, R. *180*
Kaji, E. *180*
Kajiwara, A. *180, 181*
Kakisawa, H. *179, 181*
Kamandi, K. *180*
Kameyama, M. *187*
Kanematsu, K. 111, *178*

Kanes, K.J. *187*
Karlsson, B. *180*
Karp, J.E. *185*
Karpha, A. *192*
Kato, T. 83, 84, 87, *180, 181*
Kelsey, R.G. *191*
Kende, A.S. 118, 119, *180*
Khan, N.U.-D. 7, *180*
Kim, H.B. *179*
Kingston, D.G.I. 1, 7, *176, 179, 180, 182,*
    185, 189, 191, 192
Kirfel, A. *178*
Kitagawa, I. 114–117, *180, 181, 186*
Kitamura, K. *180*
Kleine-Klausing, A. 150, *174*
Kobayashi, T. *181*
Koch, B. *178*
Kojima, T. *179, 181*
Kondo, H. *181*
Koyama, H. *186*
Krakoff, I.R. *181*
Kraus, G.A. 102, *181*
Krauss, N.E. *187*
Kris, M.S. *181*
Krol, W.J. *192*
Kühling, A. *183*
Kuhn, J. *184*
Kumagai, T. *181*
Kumar, I. *188*
Kurabayashi, M. *183*
Kurokawa, N. *186*
Kurono, M. 8, *181, 183*
Kuwajima, I. 111–113, 172, 173, *177–179,*
    181, 191

Lallemand, J. *175*
Lansing, A. *191*
Lapinjoki, S.P. *189*
Laurent, D. *177*
Lavelle, F. *174, 178*
Lecomte, J.T.J. *177*
Lee, C.L. *181, 189*
Lee, G.-H. *179*
Leete, E. 154, *181*
Legha, S.S. *181*
Le Goff, M.-T. *178, 192*
Lepistö, A.-M. *190*
Lett, R.M. *177*
Leyland-Jones, B. *188*
Liang, J.-Y. *181, 183, 191*

Lim, P. *188*
Lin, C.M. *178*
Lin, J. *175, 181*
Lin, Y.-C. *179, 182*
Lin, Y.-M. *182*
Lin, Y.-S. *190*
Lipton, R.B. *188*
Liu, A.R. *189, 192*
Liu, C.-L. *182*
Liu, J.J. *191*
Liu, L.-P. *188, 189*
Liu, X. *184*
Longnecker, S.M. *176, 182*
Look, S.A. *174, 175, 186, 189*
Lothstein, L. *179*
Lowe, M.C. *178*
Lowenthal, R.E. *179*
Lucas, H. 3, *182*
Luche, M.-J. *176*
Lupia, J.A. 142, 143, *189*
Luzzo, P. *190*
Lythgoe, B. 4, 7, 9, 32, 83, 85, 87, *173, 177, 178, 182*

MacLeod, Jr., W.D. *174*
Magee, T.V. *191*
Magnusson, G. *177, 184*
Magri, N.F. *179, 180, 182, 185*
Maki, Y. *181, 182, 188*
Mamer, O. *192*
Manchand, P.S. *175*
Manfredi, J.J. 164, *179, 182*
Mangatal, L. *176, 178, 182*
Mariel, E. *183*
Martin, S. 93, 94, *182*
Masters, S. *188*
Mathew, A.E. 64, *182, 186*
Maynard, G.D. *184*
McAlpine, J.B. *191*
McCloud, T.G. *186, 189*
McClure, T.D. *182, 191*
McGuire, W.P. *182, 185*
McLaughlin, J.L. *182*
McPhail, A.T. *188*
Mehta, G. *191*
Mejillano, M.R. *182, 191*
Mellado, W. *179, 182*
Mendez, B. *176, 184*
Mielgo, A. *184*
Miller, R.W. 7, 10, 37, *182–184*

Min, Z.-D. *181, 183*
Minamoto, K. *187*
Mishima, H. *183*
Miura, H. *186*
Miyao, A. *178*
Miyazaki, M. *183*
Mizuno, M. *181*
Molinero, A.A. 1
Monasteriios, J. *176*
Monsarrat, B. *183*
Morelli, I. 7, *183*
Morihira, K. *191*
Mucciaro, T.P. *192*
Mulhauser, M. *176*
Muller, C.L. *189*
Müller, R. *190*
Murakawa, K. *180, 186*
Muschik, G.M. *186, 189*

Naaranlahti, T. *190*
Nagaoka, H. *183*
Nakadaira, Y. *175, 181, 189*
Nakajima, Y. *178*
Nakanishi, K. 5, 80, 173, *175, 178, 181–183, 189*
Nakatsuka, T. *181*
Narayanan, V.L. *176*
Nath, J.P. *182*
Nedea, M.E. *192*
Neh, H. *183*
Neidigh, K. *192*
Neto, A.G.F. *183*
Newton, L.K. *179*
Nicolaou, K.C. 172, *191*
Nikaido, M.N. *175, 181*
Niwa, M. *181*
Norin, T. *180*
Nuydens, R. *176*

Oberhansli, P. *175*
O'Connell, J.P. *181*
O'Hanlan, K.A. *177*
Ohashi, M. *178, 181*
Ohnuma, T. *183, 188*
Ohsawa, K. *183*
Ohsuki, S. *178*
Ohtsuka, Y. 96, 98, *183, 184*
Oishi, R. *187*
Oishi, T. 96–98, *183, 184*
Ojima, I. 149, 150, 172, *184, 191*

Olesen, A.S. *179*
Onuma, S. *181*
Opie, C.T. *184*
Orr, G.A. *182*
Ovington, L. *180*
Owada, J. *179*
Oxford, A.W. *175*
Ozen, H.C. *190*

Pace, R. *190*
Padilla, J. *174*
Paietta, E. *188*
Palomo, C. *184*
Papadopoulos, N. *181*
Paquette, L.A. 7, 136–139, 168, 173, *177*, *184, 192*
Parente, R.M. *181*
Park, Y.H. *191*
Parness, J. *179, 184*
Parveen, N. 7, *180*
Parvez, M. *177*
Pascard, C. *173, 175, 178*
Patel, B.P. 107, *187*
Pattenden, G. 86, 87, 171, 173, *174, 179, 191*
Pegg, N.A. *184*
Peng, S.-M. *179*
Perumattam, J.J. *174*
Peterson, J.R. *184, 190*
Pettersson, L. *177, 184*
Phillips, J. *184*
Piccariello, T. *182*
Picot, F. *175, 185*
Pilotti, A.-M. *180*
Pisetta, A. *190*
Platt, R.V. *184*
Potier, P. 7, 22, 63, 78, 143, 145, 150, 158, *175–178, 182, 185, 192*
Poupat, C. *177*
Powell, R.G. *182–184*
Prange, T. *173–175, 190*
Prelog, V. *184*
Premachandran, U. *191*

Queneao, Y. *192*

Raber, M.N. *179, 181*
Rao, C. *184*
Ravishankar, R. 92, *189*
Rawlins, D.B. *192*

Reimer, M.L.J. *182, 191*
Riley, C. *184*
Rimoldi, J.M. 1
Ring, S. *181*
Ringel, I. *184, 187*
Rizzo, J. *184*
Robertson, J.M. *175*
Rogers, R.D. *184*
Rojas, A.C. *176, 184*
Rosenshein, N.B. *182*
Rowinsky, E.K. *176, 182, 185*
Roy, S.N. *179*
Roy, S.S. *185, 192*
Rubo, L. *189*
Runowicz, C. *177*
Russell, D.R. *174*

Sabel, W. *175*
Saha, G. *185, 192*
Sakan, K.D. 92, *185*
Sakata, S.T. *192*
Sakesena, S.K. *175*
Samaranayake, G. *176, 179, 180, 185, 192*
Sanfilippo, P. *180*
Sasaki, K. *181*
Sasloff, J. *177*
Sato, H. *181*
Sato, T. *186*
Sauriol, F. *192*
Scales, B. *173*
Schell, F.M. *185*
Schiff, P.B. *179, 181, 185*
Schnick, W. *183*
Schram, K.H. *182, 191*
Schwartz, E.L. *188*
Scott, A.I. *176*
Scrowston, R.M. *173, 177, 178*
Senilh, V. *175, 178, 185*
Serra, A.A. *176*
Seufer-Wasserthal, P. *179*
Shea, K.J. 88–90, 92, 111, 171, *185, 186, 191, 192*
Shibuya, H. *180, 181, 186*
Shimizu, K. *183*
Shiro, M. *186*
Sieburth, S.McN. 110, *186*
Simon, K.J. *178*
Slichenmyer, W.J. *192*
Smith, D.A. *185*
Smith Jr., C.R. *182–184*

Snader, K.M. 173, *179*, *188*
Snapper, M.L. *188*
Snider, B.B. 140–142, 173, *186*
Soderholm, A.-C. *180*
Sorbara, L. *179*
Sorensen, E.J. *191*
Spanton, S. *191*
Springer, J.P. *187*
Stasko, M.W. *186*, *189*
Steiert, W. *173*
Stella, V.J. *182*, *186*
Stoffersen, E. *179*
Strauman, J.J. *188*
Subrahmanyam, D. *192*
Suffness, M. 7, 173, *176*, *186*
Sumimoto, M. *180*
Sun, C.M. *191*
Sundin, S. *180*
Sunguroglu, K. *187*
Suzuki, K. *181*
Suzuki, T. *180*
Swaminathan, S. *178*
Swindell, C.S. 7, 82, 105–108, 149, 150, 157, *186*, *187*

Tachibana, S. *173*
Taga, J. *187*
Tagiliapietra, S. *190*
Takagi, A. *180*, *181*
Takagi, K. *186*
Takahashi, T. 80, *181*, *187*
Takata, T. *183*
Takayanagi, H. *180*
Takoudju, M. *175*, *187*
Tanaka, T. *181*
Taylor, D.A. 5, *187*
Taylor, H.L. *188*
Taylor, R.T. *184*
Tebbe, M.J. *188*
Teja, A.S. *179*
Tekol, Y. *187*
Tenney, D.M. *181*
Thiessen, W.E. *175*, *185*
Thigpen, T. *187*
Thomas, P.J. *181*
Thriault, R.L. *179*
Tormey, D.C. *178*
Towers, G.H.N. *173*
Trippett, S. *173*
Troops, D. *184*

Trost, B.M. 121–123, *187*, *188*
Trump, D.L. *177*, *178*, *188*
Tsai, K.H. *176*
Tsiao, C. *176*
Tsujii, S. *180*, *181*, *186*
Tsuno, K. *177*
Tucker, R.W. *184*
Tutsch, K.D. *178*

Ueda, K. *181*, *187–189*
Uyehara, T. *180*, *181*
Uyeo, S. *181*, *182*, *188*, *189*

Vance, N.C. *191*
Van Echo, D.A. *188*
Vanhaelen, M.H. *179*, *192*
Vanhaelen-Fastre, R.J. *179*, *192*
Varenne, P. *185*
Vidensek, N. *188*
Virden, C.J. *175*
Viterbo, D. *190*
Vohora, S.B. *188*
Von Hoff, D.D. *184*, *188*, *192*

Wagner, R. *182*
Wahl, A. *191*
Walker, D.G. *190*
Wall, M.E. 5, *188*
Walters, R.S. *179*
Wang, J.-S. *188*, *189*
Wang, Q.-G. *189*
Wani, M.C. 5, *188*
Weber, H. *179*
Weinandy, S. *178*
Weiss, R.B. *188*
Wender, P.A. 101, 102, 108–110, 170, 171, 173, *188*, *192*
Wertheim, M.S. *181*
Wheeler, N.C. *179*, *188*
White, J.B. *182*
Whittern, D.N. *191*
Wiernik, P.H. *177*, *188*
Willebrords, R. *176*
Williams, A.D. *179*
Williamson, J.S. *190*
Williard, P.G. *189*
Willson, J.K.V. *178*
Winkler, J.D. 131–134, 171, *188*, *189*, *192*
Winterstein, E. 4, 160, *189*
Wise, S. *186*

Witherup, K.M. *174, 175, 186, 189*
Wolff, G.-J. *178*
Wood, J.M. *179*
Woods, M.C. *175, 181, 189*
Wos, J.A. *190*
Wright, M. *175, 183, 187*
Wrixon, A.D. *176*
Wu, B. *190*

Xu, L.X. *189, 192*

Yadav, J.S. 92, 93, *189*
Yamada, Y. 102, 103, *183*
Yamamoto, Y. *180, 181, 186, 188, 189*
Yamane, K. *182*
Yang, W.L. *179, 189*
Yeh, M.-K. *179, 188, 189*
Yogai, S. *179*
Yoshizaki, F. *189*
Young, C.W. *181*

Zalkow, L.H. *176, 179*
Zamir, L.O. *192*
Zeghdoudi, R. *174, 177*
Zeheb, R. *179*
Zen, S. *180*
Zhang, P.L. *184, 190*
Zhang, Q.T. *190*
Zhang, Z.-P. *178, 180, 189, 191, 192*
Zhao, M. *184, 191, 192*
Zhao, Z.-Y. *180, 189*
Zheng, R. *178*
Zhiri, A. *179*
Zhu, Z.-Q. *189*
Zimet, A.S. *183*
Zimmerman, M. *184*
Zjawiony, J.K. *190*
Zucco, M. *191*
Zucker, P.A. 142, 143, *189*
Zydowsky, T.M. *184*
Zymalkowski, F. *180*

Subject Index

Acetic acid 65, 74–76, 145
Acetic anhydride 62
Acetone 63, 73
Acetophenone 80
2α-Acetoxyaustrospicatine 17
2α-Acetoxy-2′β-deacetylaustro-
 spicatine 17
1β-Acetoxy-5α-deacetylbaccatin I 24, 48,
 165
7-Acetylbaccatin III 74
1β-Acetylbaccatin IV 26
Acetyl chloride 77
1-Acetyl-10-deacetylbaccatin III 167
9α-Acetyl-10β-deacetylspicataxine 25
9α-Acetyl-10β-deacetylspicatine 24
5α-Acetyl-5α-decinnamoyltaxagifine 21,
 46
Acetyloxyacyl chloride 148
2′-Acetyltaxol 65
7-Acetyltaxol 164
Activity against B16 melanoma 161
Adams catalyst 74
Alkoxyacylchloride 148
Allyl chloride 83, 85
Allylic chloride 121
5α-O-(3′-Amino-3′-phenylpropionyl)-
 nicotaxine 25
Anhydrotaxininol 79
Anhydroverticillol 85
endo-Anhydroverticillol 83
Anhydroverticillol-7,8-epoxide 155
Anticancer activity 7, 162–165, 173
Antileukemic activity 81, 161, 164
Antineoplastic activity 143
Anti-ovulatory activity 161
Antitumor activity 6, 22, 81, 161–164
L-Arabinose 85
Austrospicatine 16, 42, 54

Austrotaxine 20, 42
Austrotaxus spicata 13, 16, 17, 20, 22, 24,
 25, 31, 67
threo-3-Azido-2-hydroxy-3-
 phenylpropionic acid 146
Azobis(isobutyronitrile) 69

Baccatin 5
Baccatin I 5, 23, 56, 157
Baccatin III 5, 6, 9, 10, 22, 27, 32–34,
 50, 58, 62, 63, 65, 67, 68, 70, 71, 74, 75,
 82, 96, 143, 145, 146, 149, 150, 160, 165,
 172
Baccatin IV 26, 58, 157
Baccatin V 6, 27, 32, 50, 58, 61, 67, 69
Baccatin VI 26, 49
Baccatin VII 26
Benzenepropanoic acid 8
[7-14C]Benzoic acid 173
(2R, 3S)-N-Benzoyl-O-(1-ethoxyethyl)-3-
 phenyl isoserine 145
5-Benzoyl-isopropylidenetaxinol 37
2α-Benzoyloxy-5α-cinnamoyloxy-9α,10β-
 diacetoxy-1β,13α-dihydroxy-4(20),11-
 taxadiene 18
2α-Benzoyloxy-9α,10β-diacetoxy-
 1β,5α,13α-trihydroxy-(4)20,11-
 taxadiene 14
2α-Benzoyloxy-10β,13α-diacetoxy-
 1β,5α,9α-trihydroxy-(4)20,11-
 taxadiene 14
N-Benzoyl-threo-3-phenylisoserine 147
N-Benzoyl-β-phenylisoserine
 methyl ester 5
Bicycloheptanone 142
Bicyclo[3.3.1]nonane 102
Bicyclo[2.2.2]octadienol 93
Bicyclo[5.3.1]undecadienes 101, 102

Bicyclo[5.3.1]undecene 102
Bicyclo[5.3.1]undec-1(11)-en-4-ones 104
Biological activity 7
7β,9α-Bisdeacetylaustrospicatine 16
Brevifoliol 14, 40, 54, 167
1-Bromohepta-4,6-diene 108
α-Bromo-β-hydroxycarboxylic acid 146
(2S, 3S)-2,3-Butanediol ketal 115
tert-Butylcyclohexanone-2-
    carboxylate 132
2'-(t-Butyldimethylsilyl) taxol 65

(−)-Camphor 96
d-Camphor 114, 116
Carbon dioxide 32
Carbonyldiimidazole 65
Cembrene 83, 154, 155
epi-Cembrene 154
Cephalomannine 22, 29, 33, 34, 51, 52, 58,
    61, 67–69, 165, 167
Ceric ammonium nitrate 148
Chlorobenzoate 144
N-Chloro-N-sodio-tert-butyl-
    carbamate 144
Chromic acid 71
Chromium trioxide 71
13-Cinnamoylbaccatin-III 143
Cinnamoyl chloride 144
5α-Cinnamoyloxy-2α,13α-dihydroxy-
    9α,10β-diacetoxy-4(20),11-taxadiene 18,
    40, 166
5α-Cinnamoyloxy-10β-hydroxy-2α,9α,13α-
    triacetoxy-taxa-4(20),11-diene 18
5α-Cinnamoyloxy-10β-hydroxy-2α,9α,13α-
    triacetoxy-4(20),11-taxadiene 166
5α-Cinnamoyloxy-2α,9α,10β,13α-
    tetraacetoxy-4(20),11-taxadiene 18
5α-Cinnamoyloxy-9α,10β,13α-triacetoxy-
    taxa-4(20),11-diene 18
5-Cinnamoyltaxicin-I 63, 64
O-Cinnamoyltaxicin-I 19, 67
O-Cinnamoyltaxicin-I triacetate 4, 5, 7, 11,
    19, 32, 61, 66
5-O-Cinnamoyltaxicin-I triacetate 152
O-Cinnamoyltaxicin-II triacetate 4, 7, 19
5-Cinnamoyl taxoids 18
Cisplatin 162
Claisen reaction 71

Claisen rearrangement 120, 121, 172
Cleomeolide 156
13C-NMR spectra 54–59
Colchicine 164
Comptonine 17, 42
Cotton effect 37
Cremophor EL 162
Cupric sulfate 63
(S)-(+)-trans-Cyclodecene 37
Cyclohexa-1,3-diene 80
Cyclohexanone cyanohydrin 105, 106
Cyclohexenyllithium 137, 142, 143
Cytotoxic activity 5

2'β-Deacetoxyaustrospicatine 17, 42, 54
2'-Deacetoxyaustrotaxine 20
2α-Deacetoxytaxinine J 19
2'β-Deacetylaustrospicatine 16
7β-Deacetylaustrospicatine 16
2'-Deacetylaustrotaxine 20
5α-Deacetylbaccatin I 23, 56
10-Deacetylbaccatin III 11, 22, 27, 33–35,
    62, 63, 67, 71, 75, 143, 144, 166, 168
10-Deacetylbaccatin V 67
7,9-Deactylbaccatin VI 167
10-Deacetylcephalomannine 29, 51, 52, 67
2α-Deacetyl-5α-decinnamoyltaxagifine 21,
    46
10-Deacetyl-10-oxo-7-epi-taxol 31, 51, 52
10β-Deacetylspicatine 24
4-Deacetyltaxagifine III 30
10-Deacetyltaxol 28, 67, 69, 144, 145
10-Deacetyl-7-epi-taxol 67, 69
N-Debenzoyl-N-t-butoxycarbonyl-10-
    deacetyltaxol 145
2-Debenzoyl-4,10-dideacetyl-7-
    triethylsilylbaccatin III 68
5α-Decinnamoyltaxagifine 21
Decinnamoyltaxinine J 12, 40
1-Dehydroxybaccatin III 27, 50
1-Dehydroxybaccatin IV 167
1β-Dehydroxybaccatin IV 26, 49
1β-Dehydroxybaccatin VI 26, 49
1β-Dehydroxy-4α-deacetylbaccatin IV
    49
De Mayo reaction 99, 172
N-Demethylnicaustrine 25
5-Deoxy-4,16-dihydrotaxicin I 35

5-Deoxy-4,20-dihydrotaxicin I 37
9α,10β-Diacetoxy-2α-benzoyloxy-
    5α-cinnamoyloxy-1β,13α-dihydroxy-
    4(20),11-taxadiene 167
9α,10β-Diacetoxy-2α-benzoyloxy-
    1β,5α,13α-trihydroxy-4(20)11-
    taxadiene 167
9α,10β-Diacetoxy-5α,13α-dihydroxy-4(20),
    11-taxadiene 12, 167
7β,9α-Diacetoxy-5α,13α,14β-trihydroxy-
    10-oxo-(4)20,11-taxadiene 20, 40
7,13-Diacetylbaccatin III 168
2′,7-Diacetyltaxol 51, 52
1,5-Diazabicyclo[5.4.0]undec-7-ene 65, 71
2,6-Dibromopyridine 110
Dichloromethane 33, 67, 71
Diels-Alder reaction 88, 89, 91, 92, 127
9-Dihydro-13-acetylbaccatin III 167
Dihydroquinidine 4-chlorobenzoate 147
Dihydroquinine acetate 144
Dihydrotaxinol 73
2α,13α-Dihydroxy-9α,10β-diacetoxy-5α-
    cinnamoxy-4(20),11-diene 54
2α,5α-Dihydroxy-19-hydroxymethyl-
    9α,10β,13α-triacetoxy-4(20),11-
    taxadiene 166
2α,5α-Dihydroxy-7β,9α,10β,13α-
    tetraacetoxy-(4)20,11-taxadiene 12
Dimedone 105, 106, 129
3-(Dimethylamino)-3-phenylpropanoic
    acid 4
(3R)-Dimethylamino-3-phenylpropanoic
    acid 154
4-Dimethylaminopyridine 145, 149
2,6-Dimethylbenzoquinone 103
2,6-Dimethylcyclohexanone 119
Dioxane 80
Di-2-pyridyl carbonate 145
7,13-Di(triethylsilyl) hexahydrobaccatin
    III 68
2′,7-Di(troc) taxol 65
1,2-Divinylcyclobutanols 140

Epoxy-allylsilane 86
Ethanol 32, 33, 162
Ethanolic potassium hydroxide 79
2′-(1-Ethoxyethyl)-7-(triethylsilyl)taxol 149
Ethyl acetate 33, 74
Ethyl crotonate 143

Ethyl-cis-β-phenylglycidate 146
Ethyl phenylpropiolate 172
Ethyl vinyl ether 148

Florisil 33
Formic acid 64

Geranyl bromide 87
Geranyl cyanide 83
Geranylgeranyl pyrophosphate 83, 154,
    156
Glycidic ester 147
Grignard reaction 148
Grob fragmentation 103

4,6-Heptadienal 108
2α,5α,7β,9α,10β,13α-Hexaacetoxy-(4)20,11-
    taxadiene 14
1β,2α,5α,9α,10β,13α-Hexahydroxy-
    (4)20,11-taxadiene 14, 167
Hexane 33
^{1}H-NMR spectra 40–52
Homocamphor 125
Horeau's method 8
1β-Hydroxybaccatin I 23, 48, 56
19-Hydroxybaccatin III 27, 50
10-(β-Hydroxybutyryl)-10-
    deacetylcephalomannine 29
10-(β-Hydroxybutyryl)-10-
    deacetyltaxol 29
2α-Hydroxy-2′β-deacetoxyaustro-
    picatine 17
1β-Hydroxy-7β-deacetoxy-7α-hydroxy-
    baccatin I 24, 48
1β-Hydroxy-5α-deacetylbaccatin I 23, 48,
    56
1β-Hydroxy-7,9-deacetylbaccatin I 167
14β-Hydroxy-10-deacetylbaccatin III 166
5α-Hydroxy-2α-α-methylbutyryloxy-
    7β,9α,10β-triacetoxy-(4)20,11-
    taxadiene 14
5α-Hydroxy-2α,7β,9α,10β,13α-
    tetraacetoxy-(4)20,11-taxadiene 12
1β-Hydroxy-2α,5α,7β,10β-
    tetradeacetylbaccatin I 166

3-Isobutoxycyclohex-2-enone 87
Isopropanol 33
Isoproterenol 160

Jones' reagent 71

Lithium aluminum hydride 73, 75
Lithium chloride 168

Manganese dioxide 70
Maytansine 164
McMurry cyclization 119
McMurry Ti reagent 119
Meerwein's reagent 76
Melanoma 6
Mesyl chloride 78
Methanol 33, 67, 69, 71, 74
Methanolic sodium methoxide 67
5α-O-(3'-Methylamino-3'-phenylpro-
    pionyl)nicotaxine 25
Methyl 4-bromocrotonate 141
2α-α-Methylbutyryloxy-5α,7β,9α,10β-
    tetraacetoxy-(4)20,11-taxadiene 14
2α-α-Methylbutyryloxy-5α,7β,10β-
    triacetoxy-(4)20,11-taxadiene 14
Methyl cinnamate 147
5-Methyl-1,3-cyclohexanedione 103
1-Methylcyclohexene 126
2-Methylcyclohexenol 121
3-Methylcyclohexenone 116, 118
3-Methyl-2-cyclohexen-1-one 127
3-Methylcyclohex-2-enone 115
6-Methyl-5-hepten-2-one 141
Methyl iodide 63
N-Methylmorpholine N-oxide 147, 152
Mevalonate 173
Michael reaction 117
Mitsunobu procedure 152
Monoperphthalic acid 70
Mukaiyama aldol reaction 119
Myrcene 101

Neryl phenyl sulfone 85
Nicaustrine 25, 44, 56
Nicotaxine 25, 44
Nicotinic acid 22
Nitrous acid 146
Norrish II cleavage 103
A-Nortaxoid 79
A-Nortaxol 78

Octadecyl silica 167
Octalone 123

Olefin 37
Osmium tetroxide 69, 144, 147, 152
Overhauser effect 168
13-Oxobaccatin III 50, 70
13-Oxobaccatin V 70
13-Oxo-10-deacetylbaccatin III 75
2-Oxo-7,7-dimethyl-1-
    vinylbicyclo[2.2.1]heptane 169
D-2-Oxo-7,7-dimethyl-1-vinylbicyclo-
    [2.2.1]heptane 138
(R)-2-Oxo-7,7-dimethyl-1-
    vinylbicyclo[2.2.1]heptane 169
4-Oxopimelate 123
7-Oxotaxol 72
Oxy-Cope rearrangement 104, 136, 137,
    140–143, 168

(+)-β-Patchoulene oxide 96
(−)-β-Patchoulene oxide 94, 96
5α,7β,9α,10β,13α-Pentaacetoxy-
    2α-benzoyloxy-4α,20-dihydroxytaxa-
    11(12)-ene 167
7β,9α,10β,13α,20-Pentaacetoxy-
    2α-benzoyloxy-4α,5α-dihydroxytaxa-
    11(12)-ene 167
2α,5α,9α,10β,13α-Pentaacetoxy-(4)20,11-
    taxadiene 12
5α,7β,9α,10β,13α-Pentaacetoxy-4(20),11-
    taxadiene 12
Pentadienyllithium 108
4-Pentyn-1-ol 101, 110
1-Pentynyllithium 142
Phenylalanine 154, 173
cis-β-Phenylglycidic acid 146
(S)-(+)-Phenylglycine 148
threo-Phenylisoserinamide 146
(2R, 3S)-3-Phenylisoserine 146
threo-3-Phenylisoserine 146
O-β-Phenylpropionyltaxicin-I 74
threo-Phenylserine 146
Pinene 170
Piperylene 80
Podophyllotoxin 164
Potassium bromide 146
Potassium t-butoxide 153
Potassium carbonate 147
1-Propanol 33
Pseudomonas fluorescens 146
Pyrex filter 80
Pyridine 62, 71, 149

RP56976 145

9,10-Secotaxoid 86
Selenium dioxide 71
Silica gel 33, 34, 38, 65, 71, 167
Silver cyanide 144
Silver oxide 63
Silyl ether 121
Silyl ketene acetal 121
Sodium borohydride 70, 75
Sodium hydride 152
Sodium metaperiodate 71
Sodium methoxide 76
Sorbic acid 108
Spicaledonine 31
Spicataxine 25, 44, 56
Spicatine 24, 44
Spiro[3.5]non-5-en-1-one 104
Staudinger reaction 172
Sulfuric acid 5
Swern procedure 148

Taiwanxan 20, 61
Taxaceae 3
Taxacin 21
Taxadiene derivatives 39
Taxagifine 21, 46, 61
Taxagifine III 30
Taxane 150
Taxane diterpenoids 3, 7, 8, 36, 81, 143, 165
Taxanes 7, 138
Taxicin-I 67
Taxine 3–5, 19, 154, 160
Taxine A 5, 31, 32, 62, 158, 161
Taxine B 5, 17, 42, 54, 152, 161, 167
Taxinine 4, 5, 7, 8, 19, 36, 60, 66, 71, 73, 75, 79, 80, 103, 107
Taxinine A 14
Taxinine B 19
Taxinine E 19, 165
Taxinine H 14
Taxinine J 19, 40, 165
Taxinine K 30
Taxinine L 30, 80
Taxinine M 21, 46
Taxinine derivatives 32
Taxinol 75
Taxinol tetraacetate 60

Taxoids 8–12, 16, 20–23, 25, 26, 28, 30, 32, 33, 35, 37–40, 42, 44, 46, 48–50, 53, 54, 56, 58, 62, 66–71, 73, 75, 76, 79, 82, 83, 87, 88, 91, 96, 103–105, 110, 111, 121, 124, 131, 136, 138, 153, 154, 157, 166–171
Taxol 5–9, 22, 28, 32–35, 37–39, 51–53, 58, 60–62, 64, 65, 67–72, 75–79, 81, 82, 143–151, 154, 157–173
7-*epi*-Taxol 28, 51, 52, 58, 67, 159
Taxol analogues 51
Taxotere 61, 145, 148, 163, 165, 168, 172
Taxus baccata 3–5, 11, 13, 15, 17, 19, 21–24, 26–29, 31, 32, 35, 62, 143, 154, 161, 166, 167
Taxus baccata cv. *repandens* 35
Taxus brevifolia 5, 13, 15, 21, 27, 28, 31, 32, 35, 168, 173
Taxus canadensis 167
Taxus chinensis 15, 18, 19, 21, 30, 34, 167
Taxus cuspidata 4, 13, 15, 19, 21, 30, 35, 80, 168
Taxus mairei 13, 15, 18–20, 23, 24, 26, 167
Taxus media 35
Taxus X media cv. *Hicksii* 34
Taxus sp. 3–5, 34, 35, 62, 160
Taxus wallichiana 23, 27–29, 166
Taxus yunnanensis 23, 26, 27, 165, 166
Taxusin 5, 12, 54, 61, 79, 94, 96, 119, 169
(−)-Taxusin 82, 95, 96
ent-Taxusin 96
5α,9α,10β,13α-Tetraacetoxy-4(20),11-taxadiene 12
2,5,9,10-Tetraacetyl-14-bromotaxinol 8
Tetraacetyltaxinol 53, 60
2,4,4,6-Tetrabromocyclohexa-2,5-dienone 114
Tetrabutylammonium borohydride 67, 75
Tetrabutylammonium fluoride 152
Tetracyclic hemiacetal 103
Tetrahydrofuran 75
Tetrahydrotaxinine 76
Tetrahydrotaxinine dibenzoate 37
5α,9α,10β,13α-Tetrahydroxy-4(20),11-taxadiene 12
Thioethanolamine 65
Toluene 145
Toluenesulfonic acid 73
9α,10β,13α-Triacetoxy-2α-benzoyloxy-1β,5α-dihydroxy-4(20),11-taxadiene 167

5α,7β,10β-Triacetoxy-2α-(α-
    methylbutyryloxy)-
    4(20)11-taxadiene 167
7β,9α,10β-Triacetoxy-2α,5α,13α-
    trihydroxy-(4)20,11-taxadiene 12, 40
Tri-o-biphenyl phosphite 101
Tributyltin methoxide 168
Trichloroethyl chloroformate 143, 144
Triethylamine 78
Triethyloxonium tetrafluoroborate 76
7-Triethylsilylbaccatin III 63, 145, 149
7-Triethylsilyl-10-deacetylbaccatin III 63
2'-(Triethylsilyl) taxol 65
Trimethylsilyl iodide 145
Triphenylphosphine 101
2'β,7β,9α-Trisdeacetylaustrospicatine 16
2'β,13α,14β-Trisdeacetylaustrotaxine 20
7-Troc-baccatin III 145
Tubulin 6, 161, 163–165
Tubulin-assembly activity 164

Verticillene 83, 87, 154
E,E-Verticillene 86, 87
Verticillene-7,8-epoxide 155, 156
Verticillol 154

epi-Verticillol 154
Vinblastine 164
Vincristine 164
Vinyl acetate 168
1-Vinyl-2-alkenyl-7,7-dimethyl-exo-
    norbornan-2-ols 136
Vinyllithium 140, 141
Vinyl magnesium bromide 143

Wagner-Meerwein rearrangement 138,
    169
Winterstein's acid 4, 154, 158

7-(β-Xylosyl)-cephalomannine 29
7-(β-Xylosyl)-10-deacetyl-
    cephalomannine 29
7-(β-Xylosyl)-10-deacetyltaxol 28, 165
7-(β-Xylosyl)-10-deacetyltaxol C 29
7-(β-Xylosyl)-taxol 28
7-(β-Xylosyl)-taxol C 29

Yunnanxane 165, 166

Zinc bromide 77
Zinc chloride 78